aero-aero
W9-CUR-796

PERGAMON INTERNATIONAL LIBRARY
of Science, Technology, Engineering and Social Studies
The 1000-volume original paperback library in aid of education, industrial training and the enjoyment of leisure
Publisher: Robert Maxwell M.C.

ONE-DIMENSIONAL COMPRESSIBLE FLOW

THE PERGAMON TEXTBOOK INSPECTION COPY SERVICE

An inspection copy of any book published in the Pergamon International Library will gladly be sent to academic staff without obligation for their consideration for course adoption or recommendation. Copies may be retained for a period of 60 days from receipt and returned if not suitable. When a particular title is adopted or recommended for adoption for class use and the recommendation results in a sale of 12 or more copies, the inspection copy may be retained with our compliments. If after examination the lecturer decides that the book is not suitable for adoption but would like to retain it for his personal library, then a discount of 10% is allowed on the invoiced price. The Publishers will be pleased to receive suggestions for revised editions and new titles to be published in this important International Library.

THERMODYNAMICS AND FLUID MECHANICS SERIES
General Editor: W.A. WOODS

OTHER TITLES OF INTEREST IN THE PERGAMON INTERNATIONAL LIBRARY

BENSON
Advanced Engineering Thermodynamics, 2nd Edition

BRADSHAW
Experimental Fluid Mechanics, 2nd Edition

BRADSHAW
An Introduction to Turbulence and its Measurement

BUCKINGHAM
The Laws and Applications of Thermodynamics

DIXON
Fluid Mechanics, Thermodynamics of Turbomachinery, 2nd Edition

DIXON
Worked Examples in Turbomachinery (Fluid Mechanics and Thermodynamics)

HAYWOOD
Analysis of Engineering Cycles, 2nd Edition

MORRILL
An Introduction to Equilibrium Thermodynamics

PEERLESS
Basic Fluid Mechanics

The terms of our inspection copy service apply to all the above books.
Full details of all books listed will gladly be sent upon request.

ONE-DIMENSIONAL COMPRESSIBLE FLOW

H. DANESHYAR B.Eng., M.A., Ph.D.
Lecturer and Fellow of King's College,
Cambridge University, England

PERGAMON PRESS
OXFORD · NEW YORK · TORONTO · SYDNEY · PARIS · FRANKFURT

U.K.	Pergamon Press Ltd., Headington Hill Hall, Oxford OX3 0BW, England
U.S.A.	Pergamon Press Inc., Maxwell House, Fairview Park, Elmsford, New York 10523, U.S.A.
CANADA	Pergamon of Canada Ltd., 75 The East Mall, Toronto, Ontario, Canada
AUSTRALIA	Pergamon of Canada Ltd., 75 The East Mall, Toronto, Ontario, Canada
FRANCE	Pergamon Press SARL, 24 rue des Ecoles, 75240 Paris, Cedex 05, France
WEST GERMANY	Pergamon Press GmbH, 6242 Kronberg-Taunus, Pferdstrasse 1, Frankfurt-am-Main, West Germany

Copyright © 1976 H. Daneshyar

All Rights Reserved. No part of this publication may be reproduced, stored in a retrieval system or transmitted in any form or by any means: electronic, electrostatic, magnetic tape, mechanical, photocopying, recording or otherwise, without permission in writing from the publishers

First edition 1976

Library of Congress Cataloging in Publication Data

Daneshyar, H.
One-dimensional compressible flow.

(Thermodynamics and fluid mechanics series)
(Pergamon international library of science, technology, engineering, and social studies)

Includes bibliographical references.
1. Aerodynamics. 2. Turbomachines--Aerodynamics.
3. Compressibility. I. Title.

TL574.C4D36 1976 620.1'074 76-5877
ISBN 0-08-020414-7
ISBN 0-08-020413-9 pbk.

In order to make this volume available as economically and rapidly as possible the author's typescript has been reproduced in its original form. This method unfortunately has its typographical limitations but it is hoped that they in no way distract the reader.

Printed in Great Britain by A. Wheaton & Co. Exeter

CONTENTS

PREFACE

The main purpose of this book is to provide an introduction to compressible flow. It can be used as a foundation for more advanced courses such as high speed aerodynamics, compressible flow in two and three dimensions and flow with chemical reaction. The discussion is therefore limited to the simplified case of flow in one space dimension. The book should be found suitable by final year students pursuing a degree course in mechanical engineering. It should also prove useful to the engineer in industry as a refresher or an introduction to some of the fundamental concepts.

The text has evolved from a course of lectures on gas dynamics, given by the author to the final year undergraduates at the Engineering Department of the University of Cambridge. The intention of the course has been to provide a good understanding of the physical behaviour of compressible fluid flow and an adequate appreciation of the principles behind the design of modern engines. It is assumed that the reader has a basic knowledge of thermodynamics and fluid mechanics. The main concepts required are reviewed in Chapter 1 in which the basic conservation equations for mass, momentum and energy are derived for time dependent flow through a control volume. Chapters 2, 3 and 4 provide a basis for understanding steady flow with area change, friction or heat transfer. It seemed useful to include a chapter on unsteady flow as this is becoming of increasing importance in the design of modern engines. This chapter provides an introduction to the method of characteristics for solving unsteady flow problems. The reader is encouraged to use charts and tables for performing steady flow calculations. This avoids lengthy but mathematically simple manipulations which are often required. Representation in chart form makes it easier to appreciate the variations of the fluid properties.

The material presented in this book is based on several earlier books (references 5, 7, 9). The reader is referred to these books for a more detailed description and for extensions of the subject.

The author wishes to thank Dr W A Woods for editing the book and for many helpful suggestions for improvement. He also wishes to acknowledge the permission of the Council of Senate of the University of Cambridge to reproduce a number of past examination questions, the permission of Mr K Ball to reproduce figures (4.4) and (4.11), and the permission of Dr D S Whitehead for reproducing tables 5 and 6. Thanks are also due to Miss J Doble and Miss D Ellum for typing the manuscript.

NOTATION

A	cross-sectional area of duct.
a	speed of sound.
c_p	isobaric specific heat capacity.
c_v	isochoric specific heat capacity.
C	constant.
D	hydraulic mean diameter.
E	internal energy.
F	body force per unit mass.
F_x	force in the x direction.
F_i	total force on the inside walls of the duct (internal thrust).
f	friction facter = $\tau_\omega/(\frac{1}{2}\rho V^2)$.
G	mass flow rate per unit area = ρV.
g	gravitational acceleration.
h	specific enthalpy (enthalpy per unit mass).
I	impulse function = $pA + \rho AV^2$.
k	a constant.
L	length of duct.
M	Mach number = $\frac{V}{a}$.
$\mathfrak{m}$	relative molecular mass (molecular weight).
m	mass.
$\dot{m}$	mass flow rate = ρAV.
p	static pressure.
Q	heat
q	heat transfer per unit mass.
$\dot{q}$	heat transfer per unit mass per unit time.
R	characteristic gas constant
$\mathcal{R}$	universal gas constant.
s	specific entropy (entropy per unit mass).
t	time.
T	absolute temperature.
u	specific intrinsic internal energy (the internal energy per unit mass).
V	velocity of the fluid.
v	specific volume
W	work.
x	position coordinate.
z	height above a datum.

α	angle between the body force and axis of the duct.
β	bulk modulus of compression $= \rho\left(\frac{\partial p}{\partial \rho}\right)_s$.
γ	ratio of isobaric to isochoric specific heat capacities.
μ	dynamic viscosity.
ρ	density.
τ	shear stress.

<u>subscripts</u>

max	maximum.
o	stagnation condition.
r	reference.
R	reversible.
w	wall or wave.

<u>superscripts</u>

*	conditions at M = 1.
$\dot{}$	rate of change with time.
$\bar{}$	steady flow quantities
'	small perturbation.

EDITORIAL FOREWORD

The books in the Thermodynamics and Fluid Mechanics Series are a planned set of short texts, each covering specific topics. They are now well established text-books for many Engineering degree courses and they also serve as introductory reading for Engineers in industry.

The present volume, on one-dimensional compressible flow, meets the requirements of undergraduates in mechanical engineering and other engineering students taking courses in thermodynamics and fluid mechanics.

W.A.W.
1976

CHAPTER I

CONCEPTS FROM THERMODYNAMICS AND FLUID MECHANICS

Four basic laws can be readily applied in studying the flow of compressible fluids. These fundamental laws or principles upon which all the analyses presented in this book depend directly or indirectly are

(i) The law of conservation of mass

(ii) Newton's second law of motion

(iii) The first law of thermodynamics

(iv) The second law of thermodynamics.

In any flow analysis some information about the properties of the fluid must be known. This information (such as equation of state of a perfect gas) is used in connection with the basic laws to provide maximum knowledge of the flow.

The basic laws and the property relations are usually treated in thermodynamics and fluid mechanics. Therefore only a brief review of this material is presented in this chapter.

1.1 System, control volume and control surface

A system is an arbitrary collection of matter of fixed identity. The basic laws are usually stated for a system but with fluids in motion it is simpler to think in terms of a given volume of space through which fluid flows, than it is to think in terms of a particular mass of fluid of fixed identity. This volume of space is called a control volume. The surface which bounds the control volume is called a control surface; it is always a closed surface but may be either singly or multiply connected. The adaptation of the basic laws to the flow of fluid through a control surface are comprehensively dealt with in other volumes of this series [1, 2] and the general analyses are not reproduced here. Only the simplified treatment for one-dimensional flow is given.

1.2 The one-dimensional approximation

If the rate of change of fluid properties normal to the streamline direction is negligible as compared with the rate of change along the streamlines, then the flow can be assumed to be one dimensional. For flow in ducts this means that all the fluid properties can be assumed to be uniform over any cross-section of the duct. It is known that while the static pressure may be assumed to be constant over a cross-section of the duct, velocity, temperature, etc, may vary. Hence average values must be assumed for these properties. The assumption of one-dimensional flow gives satisfactory solutions to many problems where the cross-sectional area and shape change slowly along the fluid path, the radius of curvature of the duct axis is large compared with the passage diameter, and the shape of velocity and temperature profiles are approximately unchanged from section to section along the axis of the duct. If rapid changes in cross-sectional area take place, such as at a sudden enlargement, the one-dimensional assumption cannot be used in the intermediate vicinity of the change, but can often be applied to the flow between two planes, one well upstream and the other well downstream of the disturbance. Obviously the one-dimensional approach cannot give information about the variation of fluid properties and velocity normal to the streamlines.

The basic laws for the flow through the control volume shown in Fig. 1.1 can now be developed.

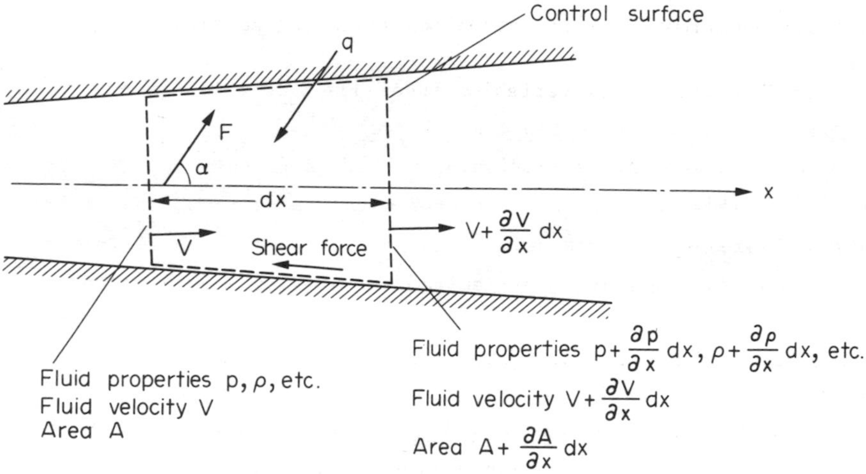

Fig. 1.1 One-dimensional flow through a control volume

For one-dimensional flow the fluid properties such as pressure, density, etc, and the particle velocity are only functions of time, t, and the distance along the axis of the duct, x (measured in the direction of flow). The variables given in Fig. (1.1) refer to a given instant of time (i.e. values obtained from a photograph taken of the flow which gives frozen values corresponding to time (t)). If the walls of the duct are rigid then the area A is only a function of the axial distance x; only flow in rigid tubes is considered in this book.

1.3 Conservation of mass: Continuity equation

The law of conservation of mass simply states that mass may neither be created nor destroyed. Thus mass of a system remains constant. For fluid flowing through the control volume of Fig. 1, the net mass flow rate into the control surface is equal to the rate of increase of mass within the control surface. At a given instant, the rate of mass flow into the control surface is ρVA and the rate of mass flow out of the control surface is $\rho VA + \frac{\partial}{\partial x}(\rho VA)\,dx$. The mass within the control surface is $\rho A\,dx$ so that the rate of change of mass of the control volume is

$$\frac{\partial}{\partial t}(\rho A\,dx)$$

Therefore $\frac{\partial}{\partial t}(\rho A\,dx) + \frac{\partial}{\partial x}(\rho VA)\,dx = 0$

or $$A\frac{\partial \rho}{\partial t} + \frac{\partial}{\partial x}(\rho VA) = 0 \qquad (1.1)$$

This is the condition for continuity of mass for one-dimensional flow in a rigid tube; equation (1.1) is therefore called the continuity equation.

For steady flow (i.e. flow variables independent of time) $\frac{\partial \rho}{\partial t} = 0$ and equation (1.1) becomes

$$\frac{\partial}{\partial x}(\rho VA) = 0 \quad \text{or}$$

$$\dot{m} = \rho VA = \text{constant} \qquad (1.2)$$

i.e. the mass flow rate ($\dot{m}$) is the same for all values of x.

1.4 Newton's second law of motion: Momentum equation

The familiar form of Newton's second law is that force exerted at a certain instant on a body is equal to the rate of change of momentum of the body at that instant. The body may be considered to be the fluid within a system, moving in the x direction. Thus the algabraic sum of the forces acting on the system in the x direction equals to the time rate of change of the momentum of the system

$$\sum F_x = \frac{d}{dt}(mV).$$

To derive a suitable relation for one-dimensional flow, consider the system shown in Fig. 1.2a, consisting of fluid in the control volume and fluid of mass δm_1 in region 1 outside the control volume. Suppose that at later time, $t + \delta t$, the system is in the position shown in Fig. 1.2b. Then for the system

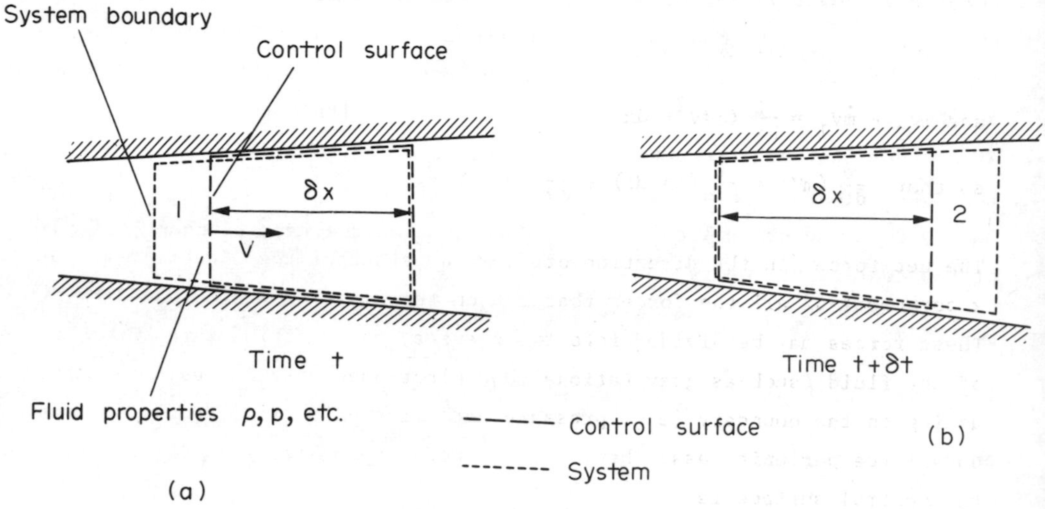

Fig. 1.2 Motion of a system through the control volume

$$\frac{d}{dt}(mV) = \lim_{\delta t \to 0} \frac{\left[(mV)_{CV} + (mV)_2\right]_{t+\delta t} - \left[(mV)_{CV} + (mV)_1\right]_t}{\delta t}$$

$$= \frac{d}{dt}(mV)_{CV} + \lim_{\delta t \to 0} \frac{\left[(mV)_2\right]_{t+\delta t} - \left[(mV)_1\right]_t}{\delta t}$$

In the limit as $\delta t \to 0$, spaces 1 and 2 coincide with the boundaries of the control volume and

$$\lim_{\delta t \to 0} \frac{(mV)_2}{\delta t} = \lim_{\delta t \to 0} \frac{\delta m_2 V_2}{\delta t} = \dot{m}_2 V_2 = \rho_2 A_2 V_2^{\,2}$$

since region 2 is now very small and the fluid in it may be considered to have uniform properties.

A similar expression holds for 1. (The term $(\dot{m}V)$ is called the flux of momentum.) Therefore

$$\frac{d}{dt}(mV) = \frac{d}{dt}(mV)_{CV} + \rho_2 A_2 V_2^{\,2} - \rho_1 A_1 V_1^{\,2}.$$

Thus for flow through the control volume the law states that the net force (in the direction of motion) acting on the fluid within the control surface is equal to the time rate of change of momentum within the control surface plus the difference between the outgoing momentum flux and the incoming momentum flux. In the limit as $\delta x \to 0$.

$$\frac{d}{dt}(mV)_{CV} = \frac{\partial}{\partial t}(\rho A\, dx) \text{ (since we are now looking at a given } x\text{)}$$

$$\text{and } \dot{m}V_2 - \dot{m}V_1 = \frac{\partial}{\partial x}(\rho AV^2)\, dx$$

$$\text{so that } \frac{d}{dt}(mV) = \frac{\partial}{\partial t}(\rho A\, dx) + \frac{\partial}{\partial x}(\rho AV^2)\, dx.$$

The net force (in the direction of flow) consists of the algabraic sum of the x component of all the forces that act on the fluid within the control surface. These forces may be divided into two classes, those acting throughout the mass of the fluid (such as gravitational and electromagnetic forces) and those acting on the boundary (e.g. pressure and friction). If F denotes the net body force per unit mass, then, the net body force acting on the fluid within the control surface is

$F\rho A\, dx$

and the component of this force in the direction of motion is

$F\rho A\, dx \cos\alpha$

where α is the angle between the force vector F and the x direction (Fig. 1.1). The forces acting on the boundary are:

(i) pressure force $= pA + p\, dA - (A + \frac{dA}{dx}dx)(p + \frac{\partial p}{\partial x}dx) = -\frac{\partial p}{\partial x}A\, dx$ in the direction of motion

(ii) shear force which results from friction.

Shear force in the direction of motion $= -\tau_\omega\, dx \times$ the wetted perimeter, where τ_ω denotes the shear stress at the wall. The shear force may be expressed in terms of the conventional pipe friction coefficient f defined by

$$f = \frac{\tau_\omega}{\frac{1}{2}\rho V^2}$$

Using the hydraulic mean diameter, D, defined as

$$D = \frac{4 \text{ x area}}{\text{wetted perimeter}}$$

$$\text{the shear force} = -\frac{\rho AV^2}{2}\, 4f\, \frac{dx}{D}.$$

Thus the application of Newton's second law gives the following momentum equation

$$-A\frac{\partial p}{\partial x} - \frac{A\rho V^2}{2}\left(\frac{4f}{D}\right) + F\rho A\cos\alpha$$

$$= \frac{\partial}{\partial t}(\rho AV) + \frac{\partial}{\partial x}(\rho AV^2) \qquad (1.3)$$

This is the general form of the momentum equation for one-dimensional flow. If the body force and the shear force are negligible, then (1.1) and (1.3) give:

$$\frac{1}{\rho}\frac{\partial p}{\partial x} + \frac{\partial V}{\partial t} + V\frac{\partial V}{\partial x} = 0 \quad \text{(Euler's equation)} \qquad (1.4)$$

For steady flow, $\frac{\partial}{\partial t}(\rho AV) = 0$ and $\frac{\partial}{\partial x}(\rho AV) = 0$ (no mass addition). Therefore (1.3) becomes

$$+\frac{1}{\rho}\frac{dp}{dx} + \frac{V^2}{2}\left(\frac{4f}{D}\right) - F\cos\alpha + V\frac{dV}{dx} = 0 \qquad (1.5)$$

which, in the absence of dissipative and body forces, gives

$$\frac{1}{\rho}dp + VdV = 0$$

For incompressible flow ρ is constant and this equation gives

$$p + \rho\frac{V^2}{2} = \text{const} \qquad (1.6)$$

If only gravitational body forces are involved then

$-F\cos\alpha = g\frac{dz}{dx}$ and equation (1.5) becomes

$$\frac{dp}{dx} + \frac{\rho V^2}{2}\left(\frac{4f}{D}\right) + \rho g\frac{dz}{dx} + \rho V\frac{dV}{dx} = 0 \qquad (1.7)$$

1.5 The first law of thermodynamics

The first law of thermodynamics expresses the conservation of energy and for a system may be written as

$$\oint đQ = \oint đW^*$$

where the integrals are taken round a cycle to the starting conditions of the system. By convention in thermodynamics heat added to a system is positive and work done on the system is negative. As a consequence of the above

---

* $\oint đQ$ and $\oint đW$ are not exact differentials because they depend on the path of the process. Following Zemansky (3) an inexact differential is denoted by a line drawn through the differential sign.

equation there exists a property of the system which is its energy content or internal energy. The first law may be written as

$$dE = đQ - đW$$

The internal energy E consists of all the different forms of energy inside the system.

$E = m(\frac{V^2}{2} + gz + u)$, where m denotes mass of the system), in the absence of other forms of energy such as electrical, magnetic, capilarity, etc.

To derive a suitable expression for one-dimensional flow, consider the system shown in Fig. 1.2a consisting of fluid in the control volume and fluid of mass δm_1 in region 1 outside the control volume. Suppose that at some time later $t + \delta t$ the system is in the position shown in Fig. 1.2b. From the first law:

$$(E_2 + E_{CV})_{t+\delta t} - (E_1 + E_{CV})_t = \delta Q - \delta W.$$

In the limit as $\delta t \to 0$, spaces 1 and 2 coincide with the boundaries of the control volume and the fluid in each region can be considered to have properties which can be only infinitesimally different from the values at the boundary.

$$\text{Therefore}\quad \dot{Q} - \dot{W} = \frac{dE_{CV}}{dt} + \lim_{\delta t \to 0} \frac{\delta m_2}{\delta t}\left[u_2 + \frac{V_2^2}{2} + gz_2\right] - \lim_{\delta t \to 0} \frac{\delta m_1}{\delta t}\left[u_1 + \frac{V_1^2}{2} + gz_1\right]$$

For $\delta t \to 0$, $\frac{\delta m_2}{\delta t} = \dot{m}_{out}$, mass flow rate out of the control volume and $\frac{\delta m_1}{\delta t} = \dot{m}_{in}$, mass flow rate into the control volume, so that

$$\dot{Q} - \dot{W} = \frac{dE_{CV}}{dt} + \dot{m}_{out}\left[u_2 + \frac{V_2^2}{2} + gz_2\right] - \dot{m}_{in}\left[u_1 + \frac{V_1^2}{2} + gz_1\right]. \qquad (1.8)$$

The rate of work done by the system $\dot{W}$ consists of several parts. At the upstream boundary work $(\dot{m}_{in}\, p_1 v_1)$ is done on the system per unit time. At the downstream boundary the system is doing work $\dot{m}_{out}\, p_2 v_2$ per unit time. Therefore equation (1.8) becomes

$$\dot{Q} - \dot{W}_o = \frac{dE_{CV}}{dt} + \dot{m}_{out}\left(h_2 + \frac{V_2^2}{2} + gz_2\right) - \dot{m}_{in}\left(h_1 + \frac{V_1^2}{2} + gz_1\right) \qquad (1.9)$$

where $\dot{W}_o$ is the rate at which (external) work crosses the control surface and $h = u + pv$ in the enthalpy. For an infinitesimal control volume, i.e. $\delta x \to 0$ equation 1.9 becomes

$$đ\dot{Q} - đ\dot{W}_o = \frac{dE_{CV}}{dt} + \dot{m}_{out}\left[(h + \frac{\partial h}{\partial x}dx) + \frac{(V + \frac{\partial V}{\partial x}dx)^2}{2} + g(z + \frac{\partial z}{\partial x}dx)\right] - \dot{m}_{in}\left[h + \frac{V^2}{2} + gz\right] \qquad (1.10)$$

and for steady flow

$$đ\dot{Q} - đ\dot{W}_o = \dot{m}\left[dh + d\frac{V^2}{2} + d(gz)\right] \qquad (1.11)$$

If $\dot{q}$ denotes rate of heat transfer per unit time per unit mass of fluid and if $\dot{W}_o$ is zero, equation (1.10) can be written as

$$\dot{q}\rho A\, dx = \frac{\partial}{\partial t}(\rho A\, dx)(u + \frac{V^2}{2} + gz)$$

$$+ \frac{\partial}{\partial x}(\rho V A)(u + \frac{p}{\rho} + \frac{V^2}{2} + gz)\, dx$$

Expanding this relation and subtracting it from the continuity equation (1.1) gives

$$\frac{D}{Dt}(u + \frac{V^2}{2}) = \dot{q} - \frac{V}{\rho}\frac{\partial p}{\partial x} - \frac{p}{\rho A}\frac{\partial}{\partial x}(AV). \qquad (1.12)$$

where $\frac{D}{Dt} = \frac{\partial}{\partial t} + V\frac{\partial}{\partial x}$

1.6 The second law of thermodynamics

The second law of thermodynamics enables us to define ideal processes and hence to specify the degree of imperfection of actual processes. There are a number of statements of the second law in classical thermodynamics, the law can be stated in the following form: "a system cannot pass through a complete cycle and produce net work while exchanging heat with a single reservoir at uniform temperature". It can be shown [(2)] that as a direct consequence of the second law the quantity $\frac{đQ}{T}$ for a reversible process is an exact differential. It is therefore a thermodynamic property. This property which is called the entropy is defined by $ds = \frac{đQ}{T}$ for a reversible process.

It can be shown that[2] for irreversible processes

$$ds > \frac{đQ}{T}. \tag{1.13}$$

It can also be shown that the rate of degradation of energy can be measured in terms of entropy changes.

An expression can be derived for entropy changes by considering a system of unit mass undergoing a reversible process in the absence of motion, gravity and other effects. In this case, with reversible heat transfer

$$Tds = đQ$$

and from the first law, with work for a reversible process,

$$đQ = du + pdv$$

so that

$$Tds = du + pdv = dh - vdp \tag{1.14}$$

Although this equation is derived for a reversible process it applies to irreversible processes also, because it is simply a relationship between the thermodynamic properties.

The second law would require that for fluid flowing through the control surface of Fig. 1.1

$$\frac{\dot{q}\rho A\ dx}{T} \leqslant \frac{\partial}{\partial t}(\rho As\ dx) + \frac{\partial}{\partial x}(\rho VAs)\ dx \tag{1.15}$$

where $\dot{q}\rho A\ \delta x$ = rate of heat transfer to the fluid in the control volume

$\frac{\partial}{\partial t}(\rho As\ \delta x)$ = rate of increase of entropy within the control volume.

$\frac{\partial}{\partial x}(\rho AVs)\ \delta x$ = entropy of fluid going out of the control volume per unit time - entropy of fluid entering the control volume per unit time. For steady one-dimensional flow between station 1 and 2

$$\rho VA\ (s_2 - s_1) \geqslant \int_{\text{control surface}} \frac{đQ}{T}$$

and if $đQ = 0$, $s_2 \geqslant s_1$.

The entropy change can be related to dissipative forces by considering the energy equation (1.11) for steady flow, i.e.

$$đQ - đW_o = dh + d\left(\frac{V^2}{2}\right) + d(gz).$$

Substituting for dh from (1.14) and assuming that external work dW_o is zero, gives

$$Tds + vdp + d(\frac{V^2}{2}) + d(gz) = đQ.$$

Introducing $ds_R = \frac{đQ}{T}$, where ds_R denotes the entropy change due to a reversible heat transfer $đQ$,

$$dp + \rho d(\frac{V^2}{2}) + \rho d(gz) + \rho T(ds - ds_R) = 0.$$

The term $\rho T(ds - ds_R)$ may be thought of as the loss of pressure resulting from irreversibilities. Comparison of the above equation with the momentum equation (1.7) gives

$$\rho T(ds - ds_R) = \tfrac{1}{2}\rho V^2(4f \frac{dx}{D})$$

so that the pressure loss due to irreversibilities is the same as the pressure loss due to friction. This result accords with expectation as irreversibilities are caused by friction in this case.

1.7 Stagnation states

From the energy equation (1.11), with no heat transfer and no external work,

$$h + \frac{V^2}{2} + gz = \text{constant} = h_o.$$

h_o is called the stagnation enthalpy; this is the enthalpy of the fluid if it is brought to rest without energy transfer. If, in addition, the flow is brought to rest isentropically, then the associated properties are referred to as stagnation properties - e.g. the pressure p_o is referred to as the stagnation pressure. Note that even though the stagnation enthalpy is constant for a flow without either heat or work transfer, the entropy of the fluid might change between two state points 1 and 2 so that the stagnation properties may be different. Fig. (1.3) illustrates the change in stagnation pressure

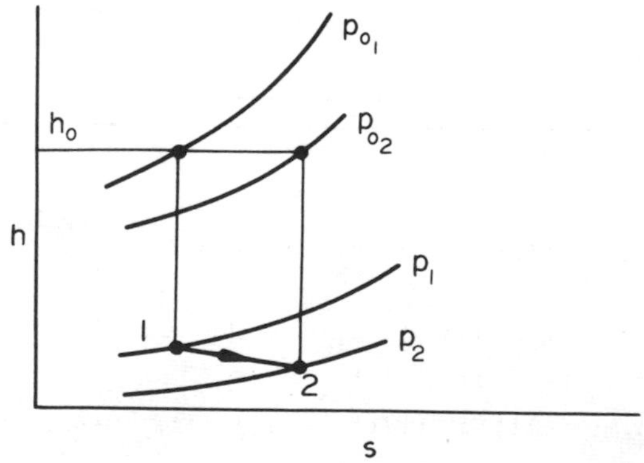

Fig. 1.3

between points 1 and 2 having the same stagnation enthalpy, but different entropies.

1.8 Compressibility and speed of sound

The fluid is said to be compressible if density changes brought about e.g. by the motion of the fluid, cannot be neglected. Some of the effects of compressibility can be appreciated by examining the velocity of propagation of an infinitesimal pressure disturbance. It will be seen that for compressible fluids the disturbance propagates with finite velocity where as if density changes are neglected the effect of the disturbance will be felt instantaneously throughout the fluid.

The velocity of propagation of the disturbance can be derived by considering a plane, infinitesimal pressure wave moving with velocity a along a duct of constant cross-sectional area, Fig. 1.4. The wave could have been generated by the piston moving from rest with a constant infinitesimal velocity δV.

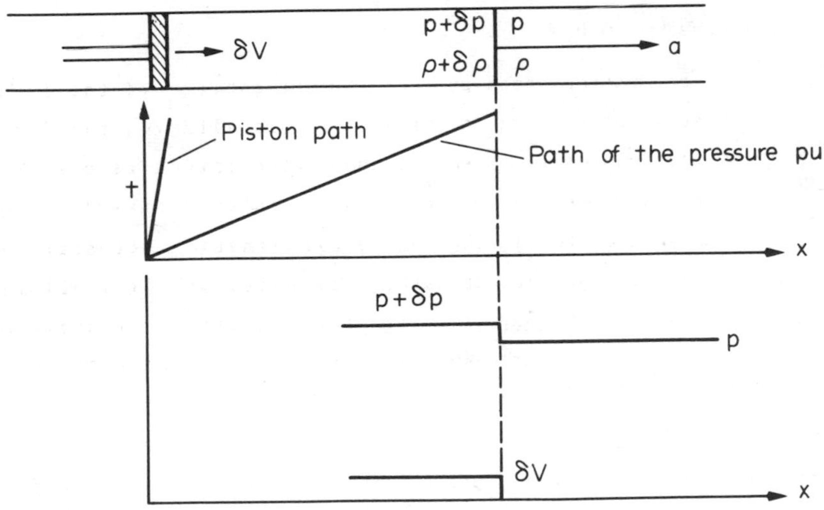

Fig. 1.4a Infinitesimal pressure pulse

The changes in pressure and density across the wave front are denoted by δp and $\delta\rho$ respectively. The flow will be steady when referred to a coordinate system moving with velocity a, which brings the pressure pulse to rest*, Fig. 1.4b. This is the flow relative to an observer moving with the pressure pulse. The fluid flows towards the observer with velocity a, pressure p and density ρ and is suddenly slowed down to velocity $(a - \delta V)$ with corresponding pressure $p + \delta p$ and density $\rho + \delta\rho$.

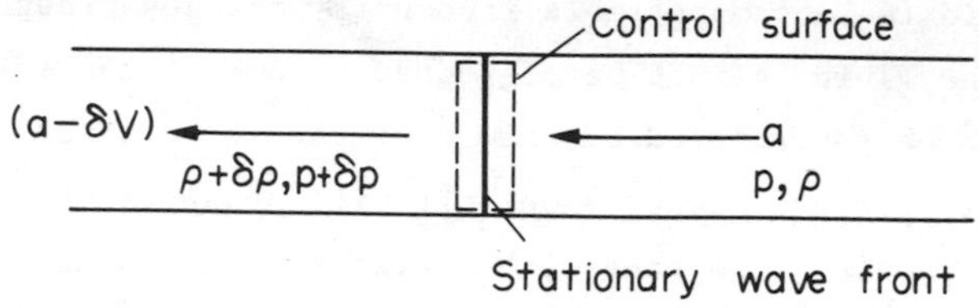

Fig. 1.4b Flow relative to an observer moving with the pressure pulse

Referring to Fig. 1.4b conservation of mass gives

$$\rho a = (\rho + \delta\rho)(a - \delta V)$$

or in the limit as δV and $\delta\rho \rightarrow 0$

$$\frac{d\rho}{\rho} = \frac{dV}{a} \tag{1.16}$$

The conservation of momentum through the wave front gives

$$p - (p + \delta p) = \rho a(a - \delta V - a)$$

so that $\delta p = \rho a\, \delta V$

and in the limit $dp = \rho a\, dV$ (1.17)

Eliminating dV from (1.16) and (1.17) gives

$$a^2 = (\frac{dp}{d\rho})$$

The value of $(\frac{dp}{d\rho})$ depends on the type of process which causes the change in the fluid properties as the fluid goes through the wave front. The process is assumed to be adiabatic, therefore the energy equation (1.11) gives

$$h + \frac{a^2}{2} = h + dh + \frac{(a - \delta V)^2}{2}$$

or in the limit $dh = a dV$

considering equations (1.17) and (1.14), $dh = \frac{dp}{\rho}$ and $Tds = 0$. The flow is therefore isentropic i.e.,

$$a^2 = (\frac{\partial p}{\partial \rho})_s \tag{1.18}$$

where the subscript s denotes constant entropy. If the fluid in the duct is

---

* If the wave front was accelerating, this procedure would not be correct. Note that in compressible flow, frames of reference are equivalent only if they move with constant velocity relative to each other, see reference 4.

flowing with finite velocity then the above analysis gives (a) as the speed of disturbances relative to the flow.

Sound waves are infinitesimal pressure disturbances and therefore equation (1.18) gives the velocity of sound in the gas within the tube. The velocity of sound is high in liquids because liquids are nearly incompressible (large pressures are required to produce small density changes). For example, for water

$$\beta = \rho\left(\frac{\partial p}{\partial \rho}\right)_s = 22.3 \times 10^5 \text{ kN/m}^2.$$

(The compressibility of a liquid is usually expressed in terms of β which is called the bulk modulus of compression.) Therefore

$$a = \sqrt{\frac{\beta}{\rho}} = 1495 \text{ m/s}.$$

For a perfect gas (i.e., a gas obeying the equation of state $\frac{p}{\rho} = RT$ and having constant specific heats) undergoing an isentropic change, equation (1.14) gives

$$dh = vdp.$$

For a perfect gas $dh = c_p\ dT = \frac{\gamma R}{\gamma - 1}\ dT$

Therefore $\frac{\gamma R}{\gamma-1}\frac{dT}{T} - \frac{dp}{\rho T} = 0,$

or $\frac{\gamma R}{\gamma-1}\frac{dT}{T} - R\frac{dp}{p} = 0,$

i.e., $\frac{p}{T^{\left(\frac{\gamma}{\gamma-1}\right)}} = \text{constant}$

Substituting for T, from $\frac{p}{\rho} = RT$, gives,

$$\frac{p}{\rho^\gamma} = \text{const.}$$

Therefore $$a = \sqrt{\left(\frac{\partial p}{\partial \rho}\right)_s} = \sqrt{\frac{\gamma p}{\rho}} = \sqrt{\gamma RT} = \sqrt{\gamma \frac{\mathcal{R}}{m} T} \qquad (1.19)$$

where $\mathcal{R}$ is the universal gas constant (= 8.3143 kJ/kmol K) and m is the relative molecular mass (molecular weight) of the gas. Thus for a perfect gas the velocity of sound depends on the relative molecular mass of the gas and on its temperature

Examples: At ambient temperature (288 K) using (1.19),

a = 91.5 m/s for Freon (Fe)

a = 1220 m/s for Hydrogen (H_2)
a = 340 m/s for air (which contains 21% oxygen (O_2) and 79% atmospheric nitrogen (N_2) by volume and has a relative molecular mass of 28.97).

It will be shown subsequently that if there are finite changes in the flow properties across the wave (such as occur in shock waves and explosion waves) then the velocity of propagation of the wave is greater than that given by (1.18).

The above analysis shows that a compressible fluid may be considered to be elastic by virtue of its variable density and the velocity of sound may be thought of as a measure of elasticity of the fluid. The larger the derivative $\left(\frac{\partial \rho}{\partial p}\right)_s$ the lower the velocity of sound. This may be extended to relatively incompressible solids where the velocity of sound is very high.

1.9 Classification of compressible flows - Mach number

There are large differences in flow patterns with compressible flows. General behaviour of the flow depends on whether the fluid velocity is greater or less than the local velocity of sound. Consider the spreading of weak disturbances produced by a line source (perpendicular to the plane of the paper) in two-dimensional flow. At any instant of time the source may be imagined to emit an infinitesimal pressure wave which spreads from the point of emission with the speed of sound relative to the fluid. If the flow velocity is smaller than the local velocity of sound, i.e., flow is subsonic, then the disturbances can be communicated throughout the flow (Fig. 1.5a). If the flow is sonic, then the disturbances form a plane front (containing the source) which separates the undisturbed fluid ahead from the disturbed fluid behind it. (Fig. 1.5b). If the flow is supersonic then the infinitesimal disturbances are swept downstream. The fronts of these disturbances form two inclined - plane surfaces, known as the Mach wedge, which represents the boundaries outside of which the fluid is undisturbed. (Fig. 1.5c)

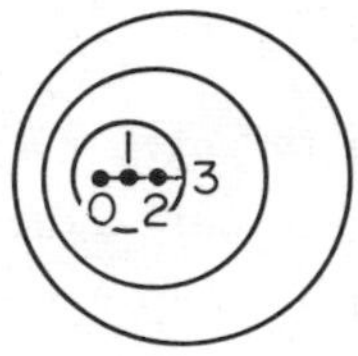

1.5(a) Subsonic flow

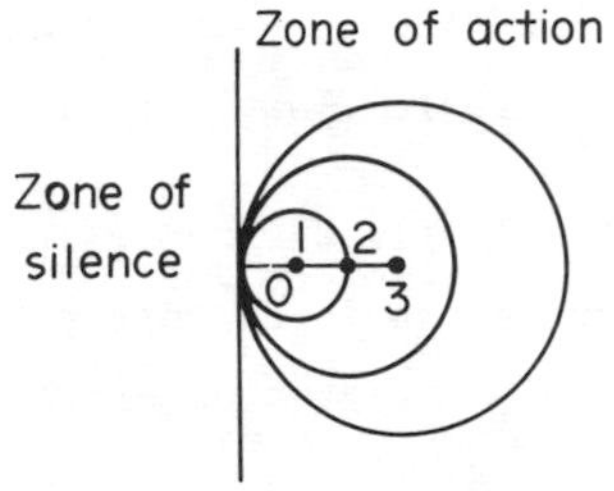

1.5(b) Sonic flow

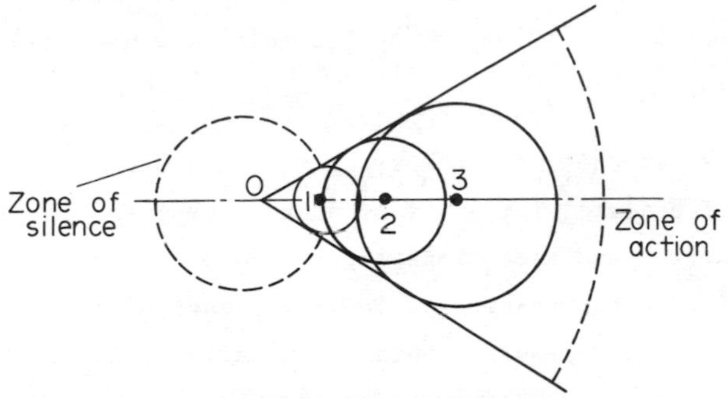

1.5(c) Supersonic flow

Fig. 1.5 Propagation of infinitesimal disturbances in subsonic, sonic and supersonic flows

Therefore the ratio of the speed of the fluid at a point to the local speed of sound at that point is a useful index for identifying the flow. This ratio is called the Mach number;

$$M = \frac{V}{a}$$

$$M = \frac{V}{\sqrt{\gamma RT}} \quad \text{for a perfect gas.} \tag{1.20}$$

(Note that (a) is the local speed of sound i.e., the speed with which infinitesimal pressure disturbances propagate at a given location measured in a coordinate system moving with the local fluid velocity. For a perfect gas (a) is a function of the local temperature only. The Mach number may be considered as the ratio of the kinetic energy of the flow to the internal energy or molecular thermal energy of the gas).

Compressible flows may be classified as follows

$M < 1$, subsonic flow,

$M = 1$, sonic flow
$M > 1$, supersonic flow.

A transonic flow is defined as a flow having regions in which the flow speed changes from subsonic to supersonic. For example, transonic flows can occur in convergent-divergent nozzles and in flow over bodies.

A hypersonic flow is a supersonic flow at high Mach numbers (often defined as a flow whose Mach number is larger than 5). Hypersonic flows are so called because they require treatment somewhat different from low Mach number supersonic flows. For example effect of dissociation may not be negligible in hypersonic flows whereas it may be neglected in low Mach number supersonic flows.

It will now be shown that Mach number is the only parameter that emerges from consideration of dynamic similarity of steady compressible flows in the absence of viscosity and heat transfer. Suppose that a specific compressible flow is prescribed with pressure p, density ρ and velocity V and it is proposed to set up another flow which is dynamically similar. For this purpose a geometrically similar model may be constructed so that the corresponding variables are $k_p p$, $k_\rho \rho$, $k_V V$ where the k's are constants. Both prototype and model are subject to Euler's equation (1.4) with $\frac{\partial V}{\partial t} = 0$ since only steady flow is being considered. For the prototype

$$dp = -\rho V\, dV$$

and for the model

$$d(k_p p) = -(k_\rho \rho)(k_V V)\, d(k_V V)$$

From these two equations

$$k_V^{\,2} = \frac{k_p}{k_\rho} \quad \text{or} \quad \frac{k_V}{\sqrt{\frac{k_p}{k_\rho}}} = 1$$

This is equivalent to saying that the two flows must have the same Mach number because

$$M_{model} = \frac{k_V V}{\sqrt{\gamma \frac{k_p p}{k_\rho \rho}}} = \frac{k_V}{\sqrt{\frac{k_p}{k_\rho}}} \quad M_{prototype} = M_{prototype}$$

Thus the Mach number, and Mach number alone, is the similarity criterion for adiabatic, non-viscous, compressible flows. (This may be expected since M can be considered as the ratio of the inertia forces to the elastic forces).

1.10 One-dimensional steady flow equations in terms of Mach number

It will be convenient to express the fundamental equations in terms of Mach number because Mach number controls the flow pattern and general behaviour of compressible flows. This will be done for steady flow of a perfect gas, for which

$$\frac{p}{\rho} = RT \tag{1.21}$$

and c_p and c_v are constants.

(i) Continuity equation

$\dot{m} = \rho AV = \text{constant}$

considering equations (1.21) and (1.20)

$$\dot{m} = \sqrt{\frac{\gamma}{R}}\ \frac{pAM}{\sqrt{T}}$$

Taking the logarithm of each side and differentiating this gives

$$\frac{dp}{p} + \frac{dA}{A} + \frac{dM}{M} - \frac{1}{2}\,\frac{dT}{T} = 0 \tag{1.22}$$

(ii) Momentum equation

Neglecting body forces, the momentum equation (1.5) becomes

$$dp + \frac{\rho}{2}\,dV^2 + \frac{\rho V^2}{2}\left(4f\,\frac{dx}{D}\right) = 0$$

Dividing by p and using (1.21) this gives

$$\frac{dp}{p} + \frac{\gamma}{2}\,\frac{dV^2}{\gamma RT} + \frac{\gamma}{2}M^2\left(4f\,\frac{dx}{D}\right) = 0$$

or

$$\frac{dp}{p} + \frac{\gamma}{2}\,dM^2 + \frac{\gamma}{2}M^2\,\frac{dT}{T} + \frac{\gamma}{2}M^2\left(4f\,\frac{dx}{D}\right) = 0 \tag{1.23}$$

(iii) Energy equation

Neglecting gravitational effects in equation (1.11) and considering the definition of Mach number, $M = \dfrac{V}{\sqrt{\gamma RT}}$

$$\frac{1}{\dot{m}}\,(đ\dot{Q} - đ\dot{W}_o) = dh_o = d\left(h + \frac{V^2(\gamma RT)}{2(\gamma RT)}\right)$$

$$= d\left(h(1+\frac{\gamma-1}{2}M^2)\right)$$

Noting that for a perfect gas $dh = c_p\, dT$

$$\frac{1}{\dot{m}\, c_p}(đ\dot{Q} - đ\dot{W}_o) = d\left[T(1+\frac{\gamma-1}{2}M^2)\right] \qquad (1.24)$$

It is also useful to express the stagnation properties in terms of Mach number.

By definition, neglecting the gravitational term

$$h_o = h + \frac{V^2}{2}$$

for a perfect gas this becomes

$$T_o = T(1+\frac{\gamma-1}{2}M^2) \qquad (1.25)$$

Since, by definition the stagnation pressure is obtained by bringing the flow to rest isentropically, then

$$\frac{p_o}{p} = \left(\frac{T_o}{T}\right)^{\frac{\gamma}{\gamma-1}}$$

and therefore $p_o = p(1+\frac{\gamma-1}{2}M^2)^{\frac{\gamma}{\gamma-1}}$

1.11 Shock waves

The analysis in section (1.8) was restricted to infinitesimal changes of the fluid properties across the wave front. If the changes are finite and abrupt, the wave is called a shock wave. The formation of shock waves is described in section (5.2.1)(see also reference 5). Shock waves occur in both steady and unsteady flows. In steady flows, with certain boundary conditions, it is not always possible for the fluid properties to vary continuously throughout the flow and stationary shock waves occur (see for example section 3.2, flow in a converging-diverging duct). In unsteady flow the steepening of compression waves leads to shock waves of varying strength (as defined by pressure change across the shock) and velocity.

The present analysis is restricted to shock waves of constant strength, and normal to the flow (normal shock), either stationary or moving with constant velocity. In the latter case the flow is steady relative to an observer moving with the wave and the analysis for the stationary case applies.

Consider the shock wave shown in Fig. 1.6, where V_1 and V_2 are fluid velocities relative to the wave front.

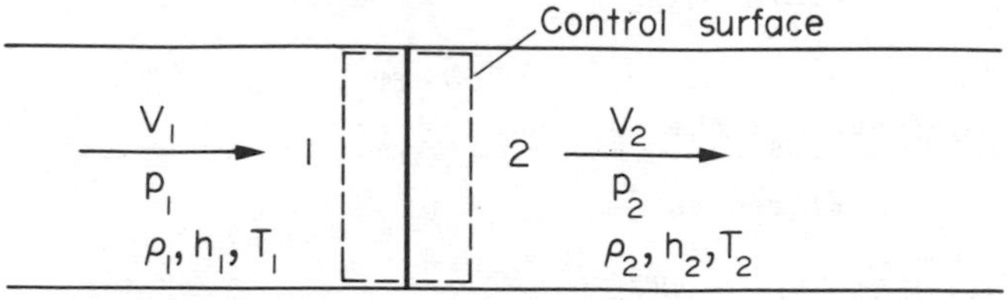

Fig. 1.6 Normal Shock

The continuity equation (1.2) gives

$$\rho_1 V_1 = \rho_2 V_2 \tag{1.26}$$

Neglecting body forces, application of Newton's second law of motion to the control surface gives;

$$p_1 + \rho_1 V_1^{\,2} = p_2 + \rho_2 V_2^{\,2} \tag{1.27}$$

or using (1.26)

$$\frac{p_1}{\rho_1 V_1} - \frac{p_2}{\rho_2 V_2} + (V_1 - V_2) = 0 \tag{1.28}$$

The flow is adiabatic and there is no external work done, therefore the energy equation (1.9) gives

$$h + \frac{V^2}{2} = h_o = \text{constant} \tag{1.29}$$

across the shock or

$$h_1 + \frac{V_1^{\,2}}{2} = h_2 + \frac{V_2^{\,2}}{2} \tag{1.30}$$

For given conditions on one side of the shock, the fundamental equations (1.26), (1.27), (1.30) provide 3 non-linear equations for the 4 unknowns p, V, h and ρ on the other side of it. One more equation is required for the solution and this is provided by the equation of state of the fluid. In general the solution of the 4 equations is not easy to accomplish and involves trial and error procedures. However for a perfect gas, analytical solution is possible.

1.11.1 Normal shock equations for a perfect gas

For a perfect gas $h_1 - h_2 = c_p(T_1 - T_2) = \frac{\gamma R}{\gamma-1}(T_1 - T_2)$ and from (1.30)

$$\frac{\gamma R}{\gamma-1} T_1 + \frac{V_1^2}{2} = \frac{\gamma R}{\gamma-1} T_2 + \frac{V_2^2}{2} = \text{constant} \tag{1.31}$$

$T = T_o$ where $V = 0$, Therefore the constant $= \dfrac{\gamma RT_o}{\gamma-1}$

Equation (1.31) can be written as

$$a_1^2 + \frac{\gamma-1}{2} V_1^2 = a_2^2 + \frac{\gamma-1}{2} V_2^2 = \gamma RT_o \tag{1.32}$$

Noting that $\frac{p}{\rho} = RT$, equation (1.28) gives

$$\frac{RT_1}{V_1} - \frac{RT_2}{V_2} + (V_1 - V_2) = 0 \tag{1.33}$$

From (1.31),

$$\frac{RT_1}{V_1} = \frac{1}{\gamma V_1}\left(\gamma RT_o - \frac{\gamma-1}{2} V_1^2\right)$$

and

$$\frac{RT_2}{V_2} = \frac{1}{\gamma V_2}\left(\gamma RT_o - \frac{\gamma-1}{2} V_2^2\right)$$

Substituting in (1.32) gives

$$\frac{1}{\gamma V_1}\left(\gamma RT_o - \frac{\gamma-1}{2} V_1^2\right) - \frac{1}{\gamma V_2}\left(\gamma RT_o - \frac{\gamma-1}{2} V_2^2\right) + (V_1 - V_2) = 0$$

or

$$\frac{\gamma RT_o(V_2 - V_1)}{V_1 V_2} + \frac{\gamma+1}{2}(V_1 - V_2) = 0$$

Therefore either $V_1 = V_2$ (trivial solution - no shock) or

$$V_1 V_2 = \frac{2\gamma RT_o}{\gamma+1} = a^{*2} \tag{1.34}$$

This relation is known as the Prandtl or the Meyer relation. The quantity a^* is the speed of sound corresponding to the state where the fluid is moving with sonic speed. (From (1.32), when $V = a$

$$a^{*2} + \frac{\gamma-1}{2} a^{*2} = \gamma RT_o \quad \text{i.e.,} \quad a^{*2} = \frac{2\gamma RT_o}{\gamma+1}\ )$$

Noting that $V = M\sqrt{\gamma RT}$ equation (1.34) can be written as

$$M_1 M_2 \sqrt{T_1 T_2} = \frac{2}{\gamma+1} T_o$$

or

$$T_o = T + \frac{V^2}{2c_p} = T \left(1 + \frac{\gamma-1}{2} M^2\right)$$

Therefore $M_1{}^2 M_2{}^2 = \left(\frac{2}{\gamma+1}\right)\left(1 + \frac{\gamma-1}{2} M_1{}^2\right)\left(1 + \frac{\gamma-1}{2} M_2{}^2\right)$

or

$$M_2{}^2 = \frac{M_1{}^2 + \frac{2}{\gamma-1}}{\frac{2\gamma}{\gamma-1} M_1{}^2 - 1} \tag{1.35}$$

This equation gives the downstream Mach number in terms of the upstream Mach number. It is also possible to express the ratios of the fluid properties $\frac{p_2}{p_1}$, $\frac{T_2}{T_1}$, etc. in terms of M_1, as follows:

Pressure ratio $\left(\frac{p_2}{p_1}\right)$

Equation (1.27) gives

$$p_1\left(1 + \frac{\rho_1}{p_1} V_1{}^2\right) = p_2\left(1 + \frac{\rho_2}{p_2} V_2{}^2\right)$$

$$\frac{p}{\rho} = RT = \frac{a^2}{\gamma} \text{ and } M = \frac{V}{a}$$

Therefore

$$\frac{p_2}{p_1} = \frac{1 + \gamma M_1{}^2}{1 + \gamma M_2{}^2}$$

Using (1.35)

$$\frac{p_2}{p_1} = \frac{2\gamma}{\gamma+1} M_1{}^2 - \frac{\gamma-1}{\gamma+1} \tag{1.36}$$

Temperature ratio $\left(\frac{T_2}{T_1}\right)$

Equation (1.31) gives

$$T_1\left(1 + \frac{\gamma-1}{2} M_1{}^2\right) = T_2\left(1 + \frac{\gamma-1}{2} M_2{}^2\right)$$

considering (1.35),

$$\frac{T_2}{T_1} = \frac{(1 + \frac{\gamma-1}{2} M_1^{\ 2})(\frac{2\gamma}{\gamma-1} M_1^{\ 2} - 1)}{\frac{(\gamma+1)^2}{2(\gamma-1)} M_1^{\ 2}} \tag{1.37}$$

Density ratio ($\frac{\rho_2}{\rho_1}$) and velocity ratio ($\frac{V_2}{V_1}$)

From (1.26) and the equation of state $\frac{p}{\rho} = RT$

$$\frac{\rho_2}{\rho_1} = \frac{V_1}{V_2} = \frac{p_2}{p_1} \times \frac{T_1}{T_2}$$

Substituting for $\frac{p_2}{p_1}$ and $\frac{T_1}{T_2}$ from (1.36) and (1.37) gives

$$\frac{\rho_2}{\rho_1} = \frac{V_1}{V_2} = \frac{\gamma+1}{2} \left(\frac{M_1^{\ 2}}{1 + \frac{\gamma-1}{2} M_1^{\ 2}} \right) \tag{1.38}$$

Entropy change across the shock

For a perfect gas equation (1.14) can be written as

$$Tds = c_p \, dT - R\frac{dp}{p}$$

Therefore

$$s_2 - s_1 = c_p \ln \left(\frac{T_2/T_1}{(p_2/p_1)^{\frac{\gamma-1}{\gamma}}} \right) \tag{1.39}$$

By definition of the stagnation pressure

$$\frac{T_o}{T} = \left(\frac{p_o}{p} \right)^{\frac{\gamma-1}{\gamma}}$$

$$\text{Therefore } \left(\frac{p_2}{p_1} \right)^{\frac{\gamma-1}{\gamma}} = \left(\frac{p_{o_2}}{p_{o_1}} \right)^{\frac{\gamma-1}{\gamma}} \frac{T_{o_1}}{T_{o_2}} \frac{T_2}{T_1}$$

For a normal shock $T_{o_1} = T_{o_2}$. Substituting for $\left(\frac{p_2}{p_1} \right)^{\frac{\gamma-1}{\gamma}}$ in (1.39) gives

$$s_2 - s_1 = - c_p \frac{\gamma-1}{\gamma} \ln \frac{p_{o_2}}{p_{o_1}} = - R \ln \frac{p_{o_2}}{p_{o_1}}$$

Also substitution from equations (1.36) and (1.37) in (1.39) gives

$$s_2 - s_1 = R \ln \left(\frac{2 + (\gamma-1)M_1^2}{(\gamma+1)M_1^2} \right)^{\frac{\gamma}{\gamma-1}} + R \ln \left(\frac{2\gamma M_1^2 - (\gamma-1)}{\gamma+1} \right)^{\frac{1}{\gamma-1}} \qquad (1.40)$$

From the second law of thermodynamics, $(s_2 - s_1) \geqslant 0$ for adiabatic flow. It can be shown that $(s_2 - s_1)$ calculated from (1.40) is positive only when $M_1 > 1$. Hence, upstream of the normal shock, flow relative to the shock must be supersonic. Also equation (1.35) shows that when $M_1 > 1$, $M_2 < 1$, i.e., downstream of the shock, flow relative to the shock must be subsonic.

<u>Pressure ratio</u> $\left(\frac{p_{o_2}}{p_1} \right)$

$$\frac{p_{o_2}}{p_1} = \left(\frac{p_{o_2}}{p_{o_1}} \right)\left(\frac{p_{o_1}}{p_1} \right) = \left(\frac{p_2}{p_1} \right)\left(\frac{T_1}{T_2} \right)^{\frac{\gamma}{\gamma-1}} \left(\frac{p_{o_1}}{p_1} \right)$$

Substitution from equations (1.36) and (1.37), and noting that for a perfect gas

$$p_o = p \left(1 + \frac{\gamma-1}{2} M^2 \right)^{\frac{\gamma}{\gamma-1}}$$

gives:

$$\frac{p_{o_2}}{p_1} = \left(\frac{\gamma+1}{2} M_1^2 \right)^{\frac{\gamma}{\gamma-1}} \left(\frac{2\gamma M_1^2}{\gamma+1} - \frac{\gamma-1}{\gamma+1} \right)^{\frac{1}{(\gamma-1)}} \qquad (1.41)$$

Values of M_2, and the temperature and pressure ratios are tabulated in table 1. Use of the table for solving practical problems is preferable to using the equations directly, since in most cases this leads to lengthy trial and error procedures.

In the above analysis M_1 was chosen as the independent variable. However it is often convenient to choose the pressure ratio $\left(\frac{p_2}{p_1}\right)$ as the independent variable. The Rankine-Hugoniot relations, express $\left(\frac{\rho_2}{\rho_1}\right)$ and $\left(\frac{T_2}{T_1}\right)$ as functions of $\left(\frac{p_2}{p_1}\right)$. These relations can be derived as follows:

Multiplying equation (1.28) by $(V_1 + V_2)$ and using equation (1.26) gives

$$V_1^2 - V_2^2 = (p_2 - p_1)\left(\frac{1}{\rho_1} + \frac{1}{\rho_2} \right) \qquad (1.42)$$

From the energy equation (1.31)

$$V_1^2 - V_2^2 = \frac{2\gamma}{\gamma-1}\left(\frac{p_2}{\rho_2} - \frac{p_1}{\rho_1}\right)$$

substituting for $V_1^2 - V_2^2$ in (1.42) gives:

$$\frac{\rho_2}{\rho_1} = \frac{1 + \frac{\gamma+1}{\gamma-1}\frac{p_2}{p_1}}{\frac{\gamma+1}{\gamma-1} + \frac{p_2}{p_1}} \tag{1.43}$$

The temperature ratio $\frac{T_2}{T_1}$ is given by

$$\frac{T_2}{T_1} = \frac{p_2}{p_1} \times \frac{\rho_1}{\rho_2}$$

Note that if the pressure change across the shock is infinitesimal i.e.,

$\frac{p_2 - p_1}{p_1} = \frac{\Delta p}{p_1} << 1$ then the above equation gives

$$\frac{\Delta\rho}{\rho_1} \simeq \frac{1}{\gamma}\frac{\Delta p}{p_1}$$

and

$$\frac{\Delta T}{T_1} \simeq \frac{\gamma-1}{\gamma}\frac{\Delta p}{p_1}$$

In the limit $\Delta p \to 0$ these relations become exact and the changes across the wave become isentropic as was shown in (1.8).

EXERCISES

1. If $G = \rho V$, $h_o = h + \frac{1}{2}V^2$, $I/A = p + \rho V^2$, then in an arbitrary gas show that

$\rho dh_o + VdG = d\,(I/A) + \rho T ds$.

Also show that

$$(M^2 - 1)\frac{dV}{V} = -\frac{dG}{G} + \frac{dh_o}{a^2} - \frac{ds}{a^2}\left(\frac{\partial h}{\partial s}\right)_\rho.$$

2. An arbitrary gas flows reversibly and adiabatically through a duct of constant cross-sectional area. If work is extracted reversibly from the gas determine the sign of the change of Mach number (a) when the flow is subsonic (b) when the flow is supersonic.

(Assume that $\left(\frac{\partial^2 p}{\partial v^2}\right)_s$ is positive and use results of question 1).

3. A perfect gas of density 1.6 kg/m^3 and pressure 68.95 kN/m^2 enters a stationary normal shock in which its velocity falls from 456 to 152 m/s. Find the density, pressure and enthalpy changes in the shock and deduce the specific heat ratio for the gas. Verify that its entropy increases in the shock and calculate the upstream and downstream Mach numbers.

Answer: 1.24, 1.98, 0.555.

4. A stream of air with a Mach number of 2.4, a static pressure of 130 kN/m^2, and a stagnation temperature of 450 K passes through a normal shock. Calculate the stagnation pressure loss, and the downstream velocity and speed of sound.

Answer: 875 kN/m^2, 217 m/s, 415 m/s.

5. A pitot tube in a supersonic wind tunnel measures a pressure of 70 kN/m^2. The static pressure upstream of the shock wave (formed due to the tube) is 15 kN/m^2. Estimate the Mach number of the tunnel.

Answer: 1.789.

CHAPTER 2

ISENTROPIC FLOW

If heat transfer may be considered negligible, and effects of friction and drag are relatively small and can be neglected, then the flow may be considered reversible and adiabatic and hence isentropic. The fluid properties can then change if there is a change in the cross-sectional area. The analysis for this case gives the ideal conditions and can be used as a standard for comparison of the actual flows.

2.1 General features of steady, inviscid isentropic flow

By using the basic equations derived in chapter 1 the general behaviour of isentropic flow can be studied.

The continuity equation 1.2 gives

$\log \rho + \log V + \log A = \text{constant}$

differentiating

$$\frac{d\rho}{\rho} + \frac{dV}{V} + \frac{dA}{A} = 0 \tag{2.1}$$

The equation of motion (1.4) with $\frac{\partial V}{\partial t} = 0$, gives

$$\frac{dp}{\rho} + V\,dV = 0 \tag{2.2}$$

Since the entropy is constant, equation 1.18 gives

$$\left(\frac{\partial p}{\partial \rho}\right)_s = \frac{dp}{d\rho} = a^2 \tag{2.3}$$

From (2.1), (2.2) and (2.3)

$$\frac{dA}{A} = -\frac{dV}{V} - \frac{d\rho}{\rho} = \frac{dp}{\rho V^2} - \frac{d\rho}{\rho} = \frac{dp}{\rho V^2}\left(1 - \frac{V^2}{\left(\frac{dp}{d\rho}\right)}\right)$$

$$= \frac{dp}{\rho V^2}\left(1 - M^2\right). \tag{2.4}$$

From (2.2) $\frac{dV}{dp} < 0$ (2.2a)

By considering equation (2.4) and (2.2a) effect of area change on the fluid properties and speed can be studied. The results are summarized in table 2.1.

TABLE 2.1

	M < 1 subsonic		M > 1 supersonic	
Area change $\frac{dA}{dx}$	$\frac{dA}{dp}$ (2.4)	pressure and velocity	$\frac{dA}{dp}$	pressure and velocity
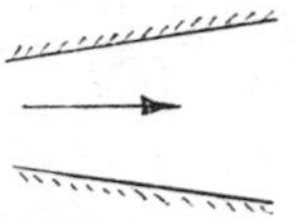 positive	> 0	p ↑ pressure increases V ↓ velocity decreases subsonic diffuser	< 0	p ↓ V ↑ supersonic nozzle
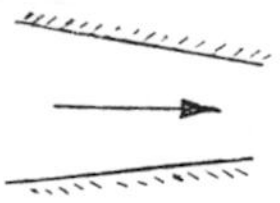 negative	< 0	p ↓ V ↑ subsonic nozzle	> 0	p ↑ V ↓ supersonic diffuser

The effects of area change on pressure and velocity are therefore opposite for subsonic and supersonic flows.

At M = 1, $\frac{dA}{dx} = 0$ (equation 2.4) i.e., area is a minimum at M = 1 regardless of the type of fluid.

2.2 Applications of isentropic flow assumption

Internal flow (e.g. flow in ducts) - Flow is not isentropic in the vicinity of the wall because of viscous forces in the wall boundary layers. If the boundary layers are thin (e.g. flow in nozzles), then the assumption usually holds good. For thick boundary layers (flow in diffusers) the assumption may give large errors. Also flow is not isentropic across shock waves, but for very weak shocks, i.e., $\frac{p_2 - p_1}{p_1} << 1$ the assumption is valid (see section 1.11.1).

External flow. This is more closely described by the assumption of isentropic flow. Non-isentropic effects are usually confined to the boundary layers on immersed bodies and across shock waves.

TABLE 2.2

Adiabatic flow of a perfect gas with no external work transfer

From (1.24) $T_o = T(1 + \frac{\gamma-1}{2} M^2) = \text{constant}$, or $\frac{dT}{T} = \frac{-d(1 + \frac{\gamma-1}{2} M^2)}{1 + \frac{\gamma-1}{2} M^2}$ (2.5)

From (1.14) $ds = \frac{dh}{T} - \frac{dp}{\rho T}$ For a perfect gas $\frac{p}{\rho} = RT$, $dh = c_p\, dT = \frac{\gamma R}{\gamma-1} dT$

Therefore $\frac{dp}{p} = \frac{\gamma}{\gamma-1}\frac{dT}{T} - \frac{ds}{R} = -\frac{\gamma}{\gamma-1}\frac{d(1 + \frac{\gamma-1}{2} M^2)}{1 + \frac{\gamma-1}{2} M^2} - \frac{ds}{R}$ (2.6)

For a perfect gas $\frac{d\rho}{\rho} = \frac{dp}{p} - \frac{dT}{T} = -\frac{1}{\gamma-1}\frac{d(1 + \frac{\gamma-1}{2} M^2)}{1 + \frac{\gamma-1}{2} M^2} - \frac{ds}{R}$ (2.7)

From definition of Mach number $\frac{dV}{V} = \frac{dM}{M} + \frac{1}{2}\frac{dT}{T} = \frac{dM}{M} - \frac{1}{2}\frac{d(1 + \frac{\gamma-1}{2} M^2)}{1 + \frac{\gamma-1}{2} M^2}$ (2.8)

From continuity (equation 1.2) $\frac{dA}{A} = -\frac{d\rho}{\rho} - \frac{dV}{V} = \frac{\gamma+1}{2(\gamma-1)}\frac{d(1 + \frac{\gamma-1}{2} M^2)}{1 + \frac{\gamma-1}{2} M^2} + \frac{ds}{R} - \frac{dM}{M}$ (2.9)

Impulse function $I = pA + \rho AV^2 = pA(1 + \frac{\rho V^2}{p}) = pA(1 + \frac{V^2}{RT}) = pA(1 + \gamma M^2)$

Therefore $\frac{dI}{I} = \frac{dp}{p} + \frac{dA}{A} + \frac{d(1 + \gamma M^2)}{1 + \gamma M^2} = -\frac{1}{2}\frac{d(1 + \frac{\gamma-1}{2} M^2)}{1 + \frac{\gamma-1}{2} M^2} - \frac{dM}{M} + \frac{d(1 + \gamma M^2)}{1 + \gamma M^2}$

(2.10)

From (1.2) mass flow rate per unit area $\frac{\dot{m}}{A} = \rho V$

Therefore $\frac{d\left(\frac{\dot{m}}{A}\right)}{\left(\frac{\dot{m}}{A}\right)} = \frac{d\rho}{\rho} + \frac{dV}{V} = -\frac{(\gamma+1)}{2(\gamma-1)}\frac{d(1 + \frac{\gamma-1}{2} M^2)}{1 + \frac{\gamma-1}{2} M^2} + \frac{dM}{M} - \frac{ds}{R}$ (2.11)

TABLE 2.3

Isentropic flow of a perfect gas with no external work transfer

From (2.5) $T_o = T(1 + \frac{\gamma-1}{2} M^2) = \text{constant}$

$$\frac{T}{T_o} = (1 + \frac{\gamma-1}{2} M^2)^{-1} \quad (2.5s)$$

$$\frac{p}{p_o} = (\frac{T}{T_o})^{\frac{\gamma}{\gamma-1}} \qquad p_o = p\,(1 + \frac{\gamma-1}{2} M^2)^{\frac{\gamma}{\gamma-1}} = \text{constant}$$

$$\frac{p}{p_o} = (1 + \frac{\gamma-1}{2} M^2)^{-\frac{\gamma}{\gamma-1}} \quad (2.6s)$$

$$\frac{\rho}{\rho_o} = (\frac{p}{p_o})^{\frac{1}{\gamma}} \qquad \rho_o = \rho(1 + \frac{\gamma-1}{2} M^2)^{\frac{1}{\gamma-1}} = \text{constant} \quad (2.7s)$$

From (2.8) $V/V^* = M^* = M \sqrt{\frac{\gamma+1}{2}} \Big/ \sqrt{1 + \frac{\gamma-1}{2} M^2}$

$$\frac{V}{\sqrt{c_p T_o}} = \sqrt{\gamma-1}\; M\,(1 + \frac{\gamma-1}{2} M^2)^{-\frac{1}{2}} \quad (2.8s)$$

$$\text{From (2.9)} \quad \frac{A}{A^*} = \frac{1}{M}\left[(\frac{2}{\gamma+1})(1 + \frac{\gamma-1}{2} M^2)\right]^{\frac{\gamma+1}{2(\gamma-1)}} \quad (2.9s)$$

$$\text{From (2.10)} \quad \frac{I}{\dot{m}\sqrt{c_p T_o}} = \frac{\sqrt{\gamma-1}}{\gamma}\;\frac{1 + \gamma M^2}{M}\;\left(1 + \frac{\gamma-1}{2} M^2\right)^{-\frac{1}{2}}$$

$$\frac{I}{I^*} = \frac{1 + \gamma M^2}{M\sqrt{2(\gamma+1)}\sqrt{1 + \frac{\gamma-1}{2} M^2}} \quad (2.10s)$$

$$\text{From (2.11)} \quad \frac{\left(\frac{\dot{m}}{A}\right)}{\left(\frac{\dot{m}}{A}\right)^*} = \frac{M(1 + \frac{\gamma-1}{2} M^2)^{\frac{-\frac{1}{2}(\gamma+1)}{(\gamma-1)}}}{\left(\frac{\gamma+1}{2}\right)^{\frac{-\frac{1}{2}(\gamma+1)}{(\gamma-1)}}}$$

$$\text{but} \quad \left(\frac{\dot{m}}{A}\right)^* = \frac{p_o}{\sqrt{c_p T_o}}\;\frac{\gamma}{\sqrt{\gamma-1}}\;\left(\frac{\gamma+1}{2}\right)^{\frac{-\frac{1}{2}(\gamma+1)}{(\gamma-1)}}$$

$$\text{hence,} \quad \frac{\dot{m}\sqrt{c_p T_o}}{A\,p_o} = \frac{\gamma}{\sqrt{\gamma-1}}\;M\left(1 + \frac{\gamma-1}{2} M^2\right)^{\frac{-\frac{1}{2}(\gamma+1)}{(\gamma-1)}} \quad (2.11s)$$

2.3 Isentropic flow of a perfect gas

The variations of the fluid properties for adiabatic flow of a perfect gas with no external work transfer (i.e., work transfer across the flow boundaries) are given in table 2.2 in terms of Mach number M and entropy change ds. Note that for two adiabatic flows with the same initial and final Mach numbers.

(i) the temperature change is the same.

(ii) Final pressure and density are always lower for the process with greater entropy increase.

(iii) Final velocities are the same.

(iv) The area must always be larger to allow passage of the same quantity of flow for the case with the greater entropy increase.

The equations involving the entropy change cannot be integrated without the specification of the change. For isentropic flow ds = 0 and all the equations may be integrated. The results are given in table 2.3. Note that stagnation states (p_o, T_o, ρ_o), are the same for all states along the duct. Therefore these quantities can be used for reference purposes. Another convenient reference state is the state at Mach number unity. An asterisk is used to denote conditions at M = 1. The ratios $\frac{V}{V^*}$, $\frac{A}{A^*}$ and $\frac{I}{I^*}$ in table 2.2 are found by integrating equations 2.8, 2.9 and 2.10. For example, from (2.8)

$$\int_{V^*}^{V} \frac{dV}{V} = \int_{1}^{M} \left(\frac{dM}{M} - \frac{1}{2} \frac{d(1 + \frac{\gamma-1}{2} M^2)}{1 + \frac{\gamma-1}{2} M^2} \right)$$

$$\text{or } \log \frac{V}{V^*} = \log \frac{M(1 + \frac{\gamma-1}{2} M^2)^{-\frac{1}{2}}}{(1 + \frac{\gamma-1}{2})^{-\frac{1}{2}}}$$

which gives (2.8s).

2.4 Choking in isentropic flow

Mass flow rate per unit area for adiabatic flow of a perfect gas is given by

$$\frac{\dot{m}}{A} = \rho V = \frac{p}{RT} M \sqrt{\gamma RT} = \sqrt{\frac{\gamma}{R}} \frac{p}{\sqrt{T}} M$$

$$= \sqrt{\frac{\gamma}{R}} \frac{p}{\sqrt{T_o}} M \sqrt{1 + \frac{\gamma-1}{2} M^2}$$

This expression can be arranged in the form of a non-dimensional parameter

$$\frac{\dot{m}\sqrt{c_p T_o}}{A\,p} = \frac{\gamma}{\sqrt{\gamma-1}}\,M\left(1+\frac{\gamma-1}{2}M^2\right)^{\frac{1}{2}} \qquad \text{(adiabatic flow)} \qquad (2.12)$$

For isentropic flow p can be expressed in terms of p_o and M (equation 2.6s) to give

$$\frac{\dot{m}\sqrt{c_p T_o}}{A\,p_o} = \frac{\gamma}{\sqrt{\gamma-1}}\,M\left(1+\frac{\gamma-1}{2}M^2\right)^{-\frac{1}{2}\frac{(\gamma+1)}{(\gamma-1)}}$$

Since $\dot{m}$ is constant and area is minimum at M = 1 then the mass flow per unit area, $\frac{\dot{m}}{A}$, is a maximum at M = 1. Thus

$$\left(\frac{\dot{m}\sqrt{c_p T_o}}{A\,p_o}\right)_{max} = \frac{\gamma}{\sqrt{\gamma-1}}\left(\frac{\gamma+1}{2}\right)^{-\frac{1}{2}\frac{\gamma+1}{\gamma-1}}$$

For $\gamma = 1.4$, $\left(\frac{\dot{m}\sqrt{c_p T_o}}{A\,p_o}\right)_{max} = 1.281$

The pressure ratio which will give this condition of maximum flow is from equation (2.6s)

$$\left(\frac{p}{p_o}\right)_{critical} = \left(1+\frac{\gamma-1}{2}\right)^{-\frac{\gamma}{\gamma-1}} = \left(\frac{2}{\gamma+1}\right)^{\frac{\gamma}{\gamma-1}},$$

$$\left(\frac{p}{p_o}\right)_{critical} = \left(\frac{p^*}{p_o}\right) = 0.528 \quad \text{for } \gamma = 1.4.$$

For the isentropic case a flow in which the fluid velocity is equal to the acoustic velocity at the section of minimum flow area is said to be choked. The phenomena of choking can be illustrated by considering the flow through a convergent nozzle, (Fig. 2.1), for different values of the reservoir pressure p_B. When p_B is slightly less than the stagnation pressure p_o (curve 2) flow is subsonic throughout the nozzle and $p_B = p_E$, where p_E is the pressure at the exit plane of the nozzle. Reducing p_B to condition 3 increases the mass flow rate. As p_B is lowered further, the mass flow rate increases until p_B/p_o reaches the critical pressure ratio (p^*/p_o), giving Mach number at the exit plane (M_E) equal to unity. If p_B is lowered from this value, the pressure p_E cannot become less than p^*, since M_E stays at 1 (section 2.1, M = 1 at cross-section of minimum area) and $p_E/p_o = p^*/p_o$ stays at

$\left(\frac{2}{\gamma+1}\right)^{\left(\frac{\gamma}{\gamma-1}\right)}$. Therefore the mass flow rate remains constant and the pressure distribution inside the nozzle remains the same as for condition 4. The nozzle is now said to be choked.

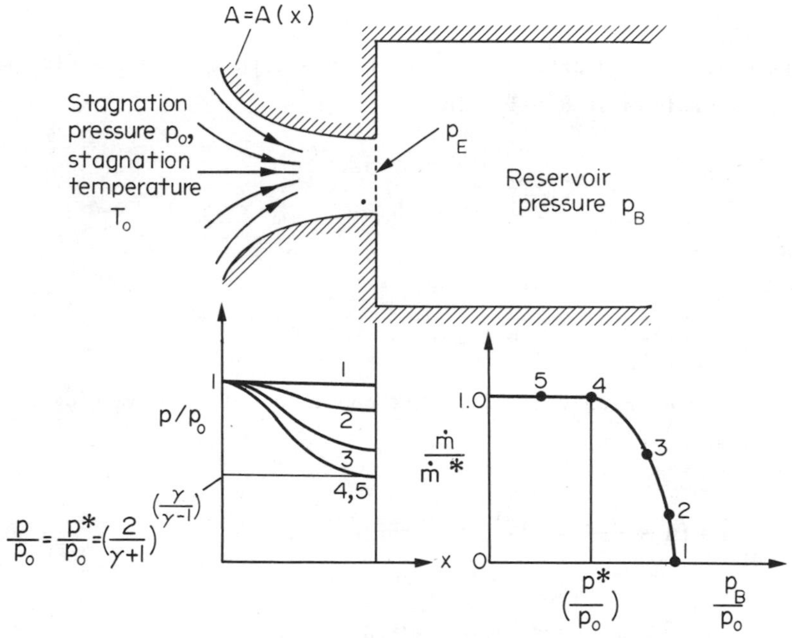

Fig. 2.1 Flow through a convergent duct for different values of the reservoir pressure (p_B).

2.5 Thrust

As the fluid flows over solid boundaries it may exert a force on the boundaries. The thrust or propulsive force is the sum of all the forces exerted by the fluid on the boundaries. All forces exerted by the solid boundaries on the fluid are countered by equal forces exerted in the opposite direction by the fluid on the solid boundaries.

The thrust produced by the fluid flowing inside a duct (internal thrust) can be found by applying Newton's second law of motion. Referring to Fig. 2.2,

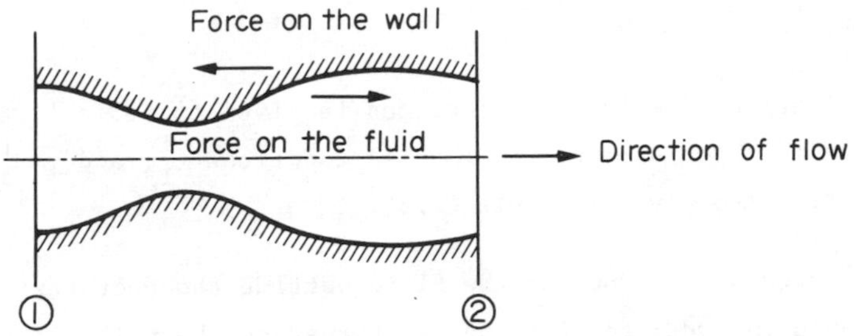

Fig. 2.2.

$$F_i + p_1A_1 - p_2A_2 = \dot{m}(V_2 - V_1) \quad (2.13)$$

where F_i denotes the total force on the walls exerted by the fluid opposite to the direction of flow (i.e. the internal thrust). Thus

$$F_i = (p_2A_2 + \rho_2A_2V_2^2) - (p_1A_1 + \rho_1A_1V_1^2) \quad (2.14)$$

For a rocket tube motor with fuel and oxidiser entering the duct at right angles to the axis (Fig. 2.3) equation (2.14) becomes

$$F_i = p_2A_2 + \rho_2A_2V_2^2 \quad (2.15)$$

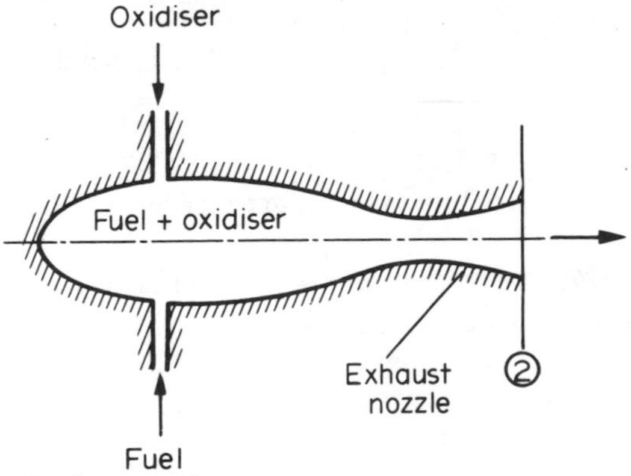

Fig. 2.3 Rocket engine.

Note that equation 2.15 assumes that the flow is steady and does not apply for transient operation of the motor.

The quantity $pA + \rho AV^2$ can be considered as a fluid impulse function. This is denoted by I.

$$I = pA + \rho AV^2$$

(Note that equation (2.14) is independent of the nature of the fluid and applies to flow with friction, heat transfer, area change, chemical reaction etc. Force F_i is due to frictional and pressure forces acting on the walls of the duct).

The impulse function (I) for a perfect gas is given in table (2.2) (i.e. $I = pA(1 + \gamma M^2)$). Non-dimensional impulse functions ($\frac{I}{I^*}$ and $\frac{I}{\dot{m}\sqrt{c_p T_o}}$) for a perfect gas are also given in table (2.2).

The force exerted on the duct by the fluid outside the duct (external thrust) should be found in order to find the net thrust on the walls. If the fluid outside the duct is stationary then this force arises from pressure forces acting on the outside wall of the duct.

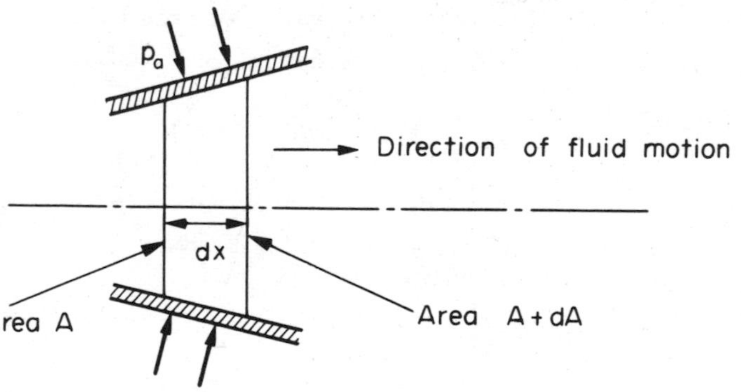

Fig. 2.4 External pressure force on an element of the duct.

Referring to Fig. 2.4, the force in the direction of motion is $p_a dA$ so that the total external thrust is

$$-\int_1^2 p_a dA = -p_a(A_2 - A_1) \qquad (2.16)$$

if p_a is constant.

From equations (2.14) and (2.16) the net force on the walls in the opposite direction of motion is

$$(I_2 - I_1) - p_a(A_2 - A_1) \qquad (2.17)$$

In the case of the rocket tube motor (Fig. 2.3)

Net thrust = $I_2 - p_a A_2 = \rho_2 A_2 V_2^2 + A_2(p_2 - p_a)$. If pressure at the exit plane is atmospheric, i.e., for subsonic flow, then $p_2 = p_a$ and

$$\text{Net thrust} = \rho_2 A_2 V_2^2 \tag{2.18}$$

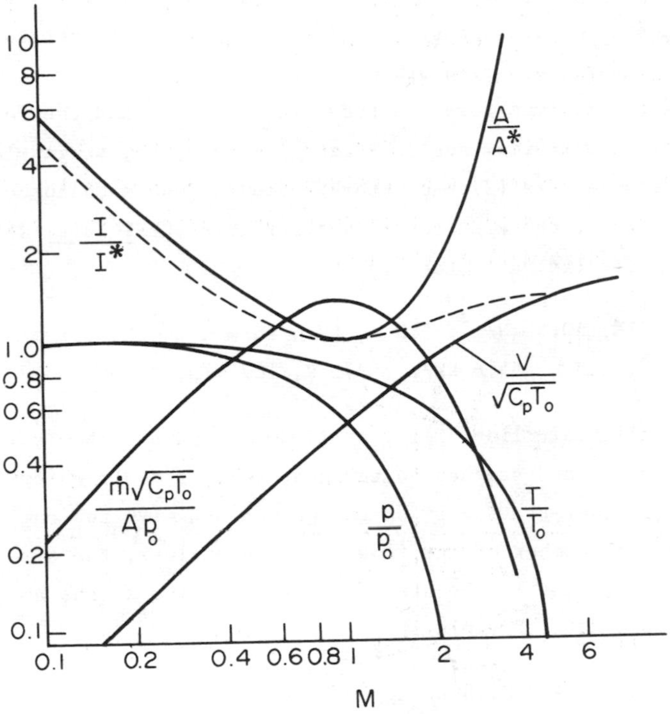

Fig 2.5 Isentropic flow functions (γ = 1.4)

2.6 Tables for isentropic flow

Each of the ratios ($\frac{T}{T_o}$, $\frac{p}{p_o}$, etc.) given in table 2.3 is only a function of Mach number and γ and can be tabulated for specified values of Mach number and γ (Table 2). The ratios can also be given in graphical form (Fig. 2.5). The tables and graphs can be used for engineering calculations and their use is usually preferred to using the equations directly because in most cases solution of the equations is tedious. (For example, if $\frac{A}{A^*}$ is given and the corresponding Mach number is required, then equation (2.9s) must be solved for M by trial and error whereas M can be read off directly from Fig. 2.5 or found from table 2). Linear interpolation is usually adequate for finding values between adjacent Mach numbers in the tables.

EXERCISES

1. At a certain point in a tube, air flows steadily with a velocity of 150 m/s and has a static pressure of 70 kN/m^2 and a static temperature of 4^oC. The flow is adiabatic and reversible.

(a) Calculate the maximum possible reduction in area, and the following quantities for that minimum area: Stagnation pressure, stagnation temperature, static pressure, static temperature, velocity and Mach number.

(b) At a point where the area is 15% smaller than the initial area, calculate the quantities listed in (a).

Answer: (a) 31%, 80.5 kN/m^2, 288 K, 42.5 kN/m^2, 240 K, 311 m/s, 1.
(b) 80.5 kN/m^2, 288 K, 64.6 kN/m^2, 271 K, 187 m/s, 0.566.

2. A jet propelled air liner flies at 15,000 m and at 265 m/s. What is the stagnation pressure and stagnation temperature of the air entering the engine? If with the air liner stationary on the ground and with the engines running with the same Mach number of the flow into the engines, the air flow to each engine is 40 kg/s, what is the airflow to each engine at the above flight conditions? (use data of table 5).

Answer: 20.4 kN/m^2, 252 K, 8.65 kg/s.

3. In a jet engine running on a test bed, the stagnation pressure and stagnation temperature just upstream of the convergent nozzle are measured to be 70 kN/m^2 gauge and 700^oC. The diameter of the nozzle exit is 0.5 m. Calculate the mass flow, jet exit velocity, and thrust of the engine. (Bar = 101.3 kN/m^2).

Answer: 45.6 kg/s, 492 m/s, 22.45 kN.

4. Consider the steady reversible adiabatic flow of a highly compressible fluid, having a pressure-density relation given by:

$$\left[\rho\left(\frac{\partial p}{\partial \rho}\right)\right]_s = \beta$$

where β is a constant. Show that

$$\frac{p}{p_o} = 1 + \frac{\beta}{p_o} \mathrm{Log}_e \left(1 - \tfrac{1}{2} M^2\right)$$

and

$$\frac{\dot{m}}{A \sqrt{\rho_o \beta}} = M \sqrt{1 - \tfrac{1}{2} M^2}$$

CHAPTER 3

FLOW IN A DUCT OF VARIABLE CROSS-SECTIONAL AREA

The effect of area variation on isentropic flow in ducts was briefly discussed in chapter 2. A more detailed study can now be made by considering flow through a convergent or a convergent-divergent duct with varying back pressure.

3.1 Convergent duct

Consider the configuration shown in Fig. 3.1, (which is similar to that of references 5,6) where a simple convergent passage discharges into a region where the back pressure p_B is controlled by a valve. Let p_E be the pressure in the exit plane of the duct and p_o the reservoir pressure (stagnation pressure for isentropic flow).

If $p_B/p_o = 1$ (condition i, Fig. 3.1) then the pressure is constant throughout the duct and there is no flow. If p_B is slightly reduced from this value (condition ii, Fig. 3.1) a subsonic flow will be established with the exit pressure p_E equal to the back pressure p_B. If p_B is reduced further to condition iii, the flow remains subsonic with $p_B = p_E$, but the mass flow rate increases. This increase continues until p_B/p_o reaches the critical pressure ratio ($p_B/p_o = p^*/p_o = p_E/p_o$) where $M_E = 1$, i.e. condition iv. If p_B is reduced further, (condition v), the pressure p_E can not become less than p^* since M_E stays at 1 (section 2.1). Therefore the flow rate remains constant and the pressure distribution inside the nozzle remains the same as for condition iv. The pressure distribution in the chamber outside the nozzle for p_B/p_E can not be predicted accurately by a one-dimensional model and is indicated by a wavy line.

The maximum mass flow rate ($\dot{m}_{max}$) can be found from equation (2.11s) if flow from the resevoir to the exit plane is isentropic. In this case, equation (2.11s), with $M_E = 1$, gives

$$\left(\frac{\dot{m}}{A}\right)_{max} = \sqrt{\frac{\gamma}{R}\left(\frac{2}{\gamma+1}\right)^{\frac{\gamma+1}{\gamma-1}}}\ \frac{p_o}{\sqrt{T_o}} \qquad (3.1)$$

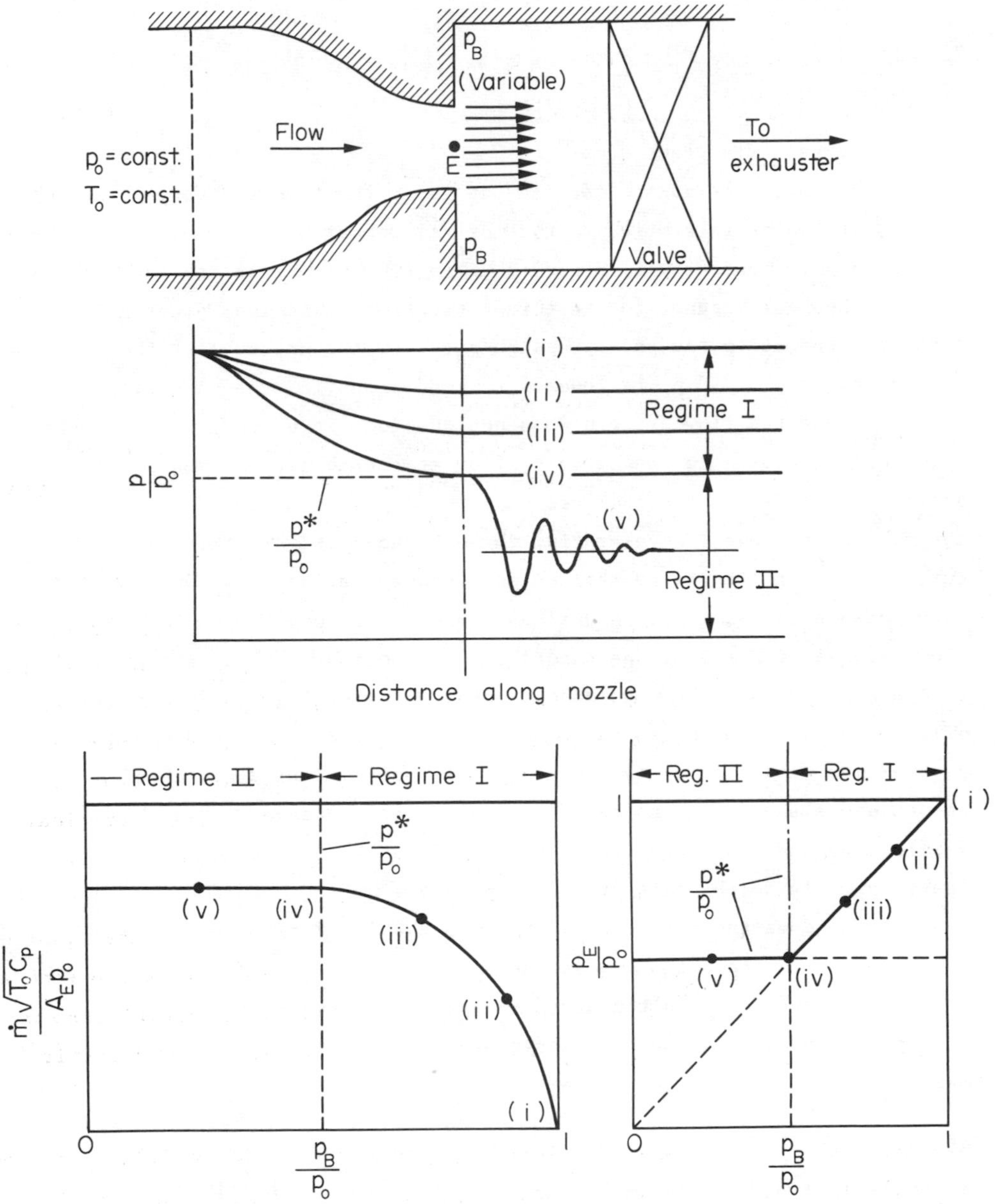

Fig. 3.1 Flow through a convergent duct.

and from equation (2.6s)

$$\frac{p^*}{p_o} = \left(\frac{2}{\gamma+1}\right)^{\left(\frac{\gamma}{\gamma-1}\right)} \qquad (3.2)$$

3.2 Convergent-divergent duct

The above experiment can be carried out for a convergent-divergent duct (Fig. 3.2) (which is similar to that of references 5,7). When p_B is slightly less than p_o, (condition 1, Fig. 3.2), the flow is similar to that in a venturi, the gas expands to the throat (section of minimum area) and diffuses after the throat to the back pressure p_B; also the pressure at the exit plane p_E is equal to p_B. If p_B is lowered to condition (1a), the mass flow rate increases but the flow pattern does not change. If p_B is lowered to the value corresponding to condition 2, flow at the throat becomes sonic but again it diffuses to the back pressure $p_B = p_E$. Condition 6 corresponds to the flow continuously accelerating through the duct. In this case the pressure ratio p_E/p_o is called the design pressure ratio of the duct. The mass flow rate is the same for conditions 2 and 6 and corresponds to the maximum flow rate since M = 1 at the throat for both cases. The flow corresponding to the back pressure between conditions 2 and 6 of Fig. (3.2) involves shock discontinuities. As p_B is reduced below the value corresponding to condition 2, the flow will accelerate supersonically for a short distance after the throat and a shock will appear in the divergent passage. This shock decelerates the flow to a higher pressure and the flow then decelerates isentropically to the exit pressure, $p_B = p_E$. As p_B is further reduced the shock moves down the nozzle until $p_B = p_4$, the limiting case of regime II, where the shock appears remove at the exit from the nozzle and the flow is everywhere supersonic in the divergent part. In regime II, the pressure $p_E = p_B$ and mass flow rate is everywhere constant and equal to the maximum mass flow rate through the passage.

As regime III is entered by reducing p_B below p_4, the duct exit pressure stays at p_6 and the flow is everywhere supersonic in the divergent part of the duct. The pressure increase from p_E to p_B occurs through oblique shock waves formed in the reservoir around the exit of the duct; flow in this region can not be treated one-dimensionally. However the shock pattern is similar to that sketched on the diagram. A well defined free streamline is formed in the reservoir and this is often observed in experiments.

For condition 6, there are no shocks anywhere in the flow and $p_B = p_E$. The exit pressure for this case is the "design" exit pressure of the duct. When p_B drops below this value (regime IV), oblique expansion waves are formed in the reservoir around the exit of the duct and a free streamline is formed; again flow in this region is not amenable to one-dimensional treatment.

In both regimes III and IV the flow pattern in the duct is independent of the back pressure p_B and is the same as for the design condition. All adjustments to back pressure are made in the reservoir near the duct exit. Thus the pressure ratio p_E/p_o in regimes III and IV depends only on the ratio of the throat area to the exit area of the duct. The pressure ratio p_E/p_o can be calculated by noting that

$M_T = 1$, therefore $\frac{\dot{m}\sqrt{c_p T_o}}{A_T p_o} = \bar{\dot{m}}_T$ can be obtained from the isentropic flow tables (table 2). $\frac{\dot{m}\sqrt{c_p T_o}}{A_E p_o} = \bar{\dot{m}}_T \times \frac{A_T}{A_E} = \bar{\dot{m}}_E$, hence p_E/p_o can be found from the isentropic flow tables. In the subsonic regime I and partly subsonic regime II, $p_E/p_o = p_B/p_o$.

The duct is choked in regimes II, III and IV and the mass flow rate, $\dot{m}$, is independent of the back pressure and is a maximum. Only in regime I can the mass flow rate be changed by variation in the back pressure.

Summarising, there are four regimes of flow.

I - Subsonic flow throughout the duct, maximum velocity is reached at the throat.

II - Subsonic flow to the throat, then supersonic up to the normal shock, followed by subsonic compression.

III - Subsonic flow to the throat, followed by supersonic flow to the exit plane - non-isentropic re-compression outside the duct through oblique shock waves.

IV - Flow in the duct identical to III, supersonic jet expanding out of the duct exit.

In practice these flow patterns are distorted by boundary layers and by two and three-dimensional effects. Because of the interaction between shock waves and boundary layers a simple normal shock is not likely to appear in a real flow. However, unless the flow separates, the one-dimensional approach will

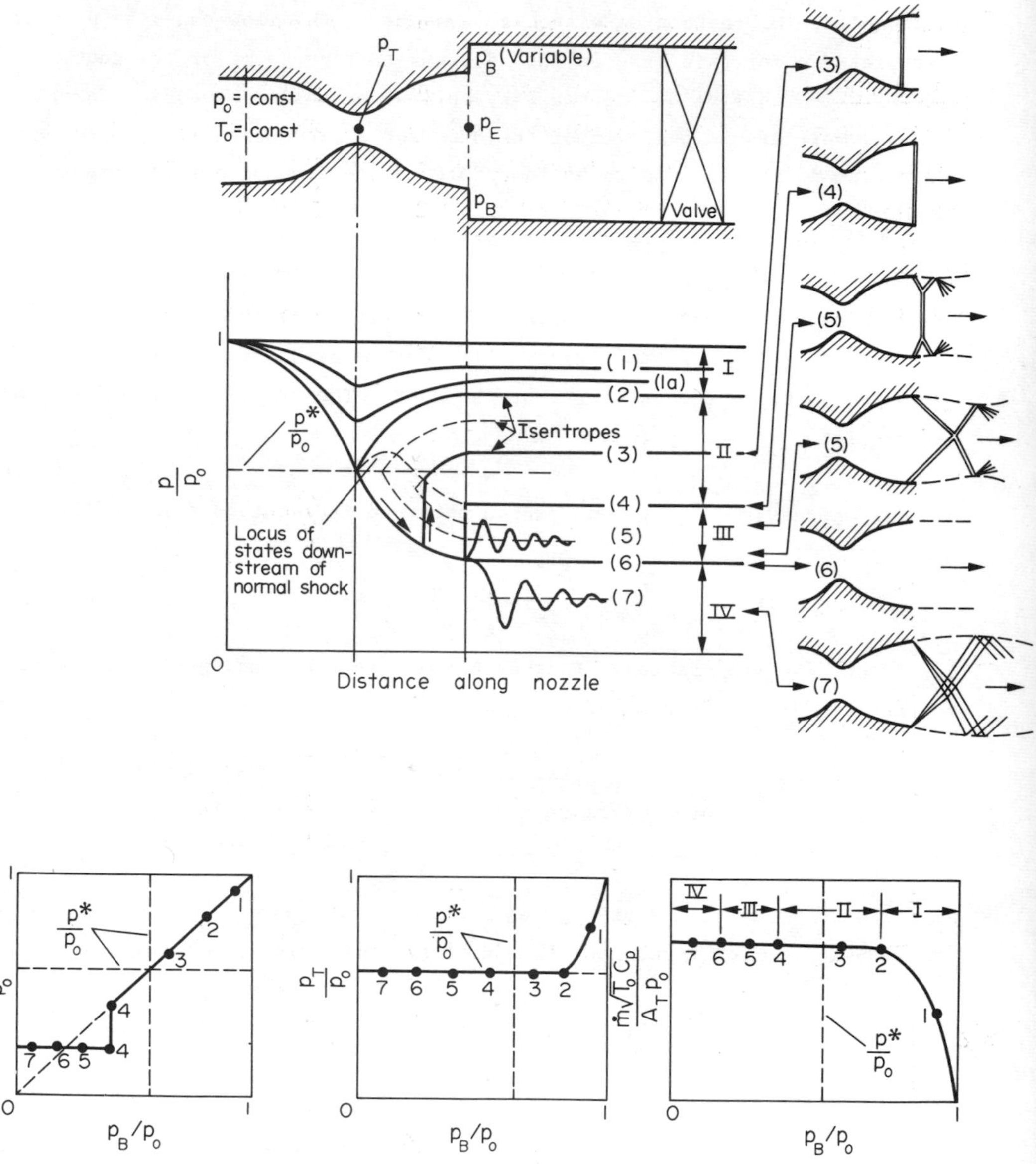

Fig. 3.2 Flow through a convergent-divergent duct.

give a reasonable approximation to the actual flow conditions.

3.3 Calculation of flow in ducts of varying cross-sectional area

In regions where flow can be assumed to be isentropic, the pressure distribution can be found from equation (2.11s) or from Fig. (3.3) which shows the variation of the mass flow function $\frac{\dot{m}\sqrt{c_p T_o}}{A\,p_o}$ with $\frac{p_o}{p}$ (as given by (2.11s)). The mass flow function increases from 0 at $\frac{p_o}{p} = 1$, reaches a maximum at M = 1 and then decreases. For given upstream conditions this is effectively a curve of $\dot{m}/A$ - (or for one test $\frac{1}{A}$ -) against $\frac{p_o}{p}$.

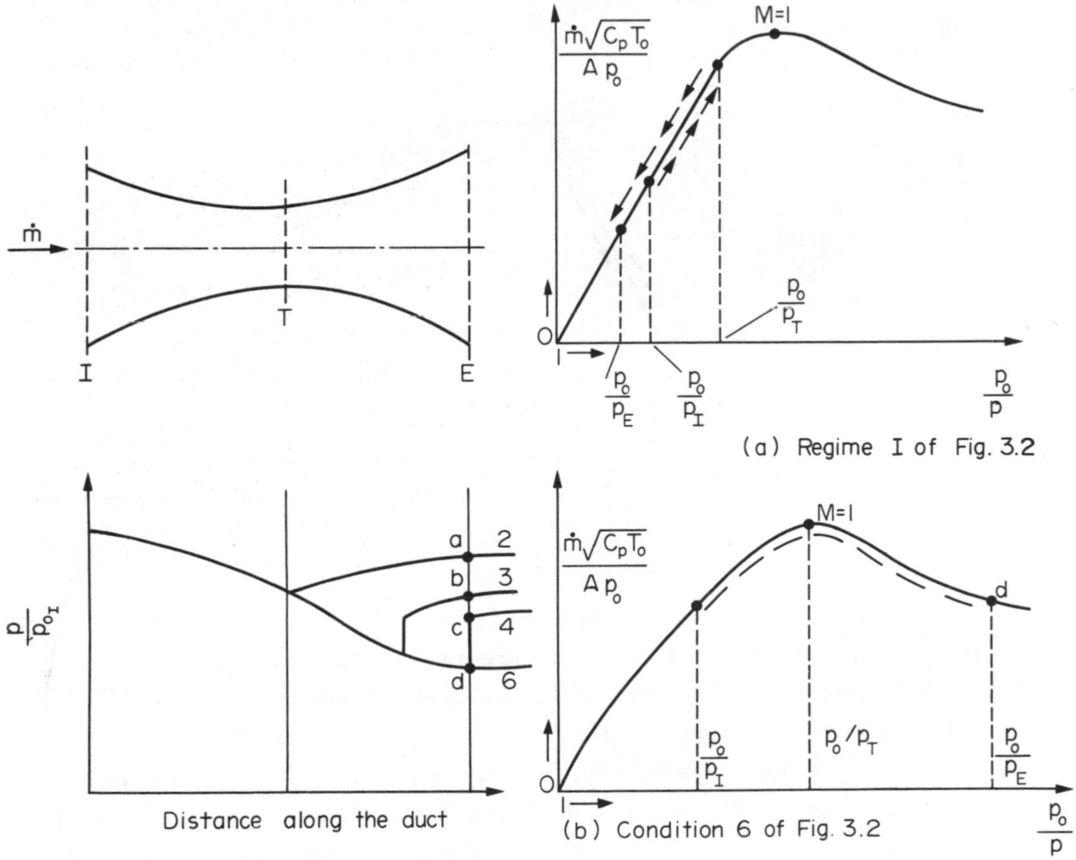

Fig. 3.3 Isentropic flow in a convergent-divergent duct.

The solution for subsonic flow is shown on Fig. (3.3a). However, if M = 1 at the throat there are two possible solutions (depending on the back pressure), the supersonic solution Fig. (3.3b), requires a much higher pressure ratio (p_o/p_E). It should be noted that for different values of the pressure ratio p_o/p_E there are a number of subsonic solutions with the limit at M = 1 but only a single supersonic solution. There is a wide range of back pressures (conditions (2) to (6) of Fig. 3.2) for which no isentropic solution is possible and for these conditions non-isentropic solutions must be sought. If a normal shock is formed in the divergent part of the duct, then the entropy of the fluid increases as it is compressed through the shock and p_o decreases. On the graphs $\frac{\dot{m}\sqrt{c_p T_o}}{A\,p_o}$ will increase from supersonic to a subsonic value. (Fig. 3.4).

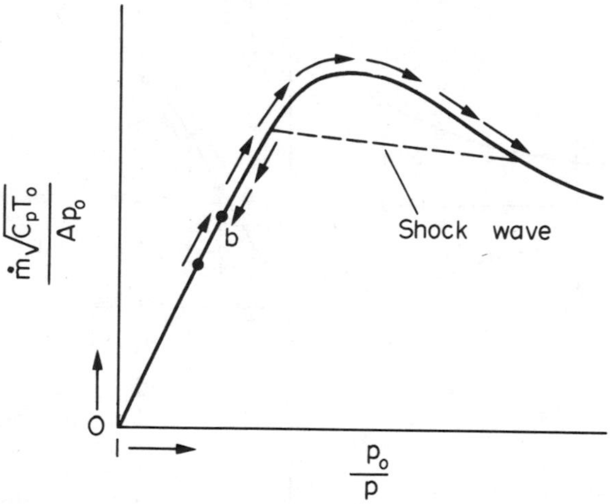

Fig. 3.4 Shock wave on the mass flow chart.

In this case the gas will have a higher pressure at the exit plane than for condition (6), (condition 3, Fig. 3.3). Allowance for the normal shock gives solutions between conditions (2) and (4); A weak shock near the throat where the Mach number is just greater than 1 will require a back pressure slightly lower than that for condition (2). A strong shock standing at the duct exit, where the Mach number is maximum will give a much greater loss of stagnation pressure and will require a much smaller back pressure than for condition (2).

3.3.1 Flow calculation in a convergent-divergent duct with a normal shock in the divergent section

Since the flow is choked, the mass flow function $\frac{\dot{m}\sqrt{c_p T_o}}{A\, p_o}$ is known at the throat. At exit the static pressure ($p_E = p_B$) and the area A_E may be given. Hence using the throat area A_T and upstream reservoir pressure p_o, the function $\frac{\dot{m}\sqrt{c_p T_o}}{A_E\, P_E}$ can be calculated. For given γ, this function depends only on Mach number and hence the Mach number at exit, M_E, can be found. Knowing M_E, the exit pressure ratio (p_E/p_{oE}) can be obtained. The total pressure ratio across the shock (p_{oE}/p_o) can now be found by noting that the total pressure is constant for isentropic flow behind the shock. This gives the area appropriate to upstream and downstream Mach number of the shock, from the normal shock and isentropic flow tables (tables 1 and 2).

Example:

A convergent-divergent duct with a throat area of half the exit area is supplied with gas ($\gamma = 1.4$) at a stagnation pressure of 140 kN/m^2. It discharges into an atmosphere of static pressure 100 kN/m^2. Show that there is a normal shock in the duct. Find the stagnation pressure of the gas at exit from the nozzle, the Mach numbers immediately upstream and downstream of the shock, and the ratio of the area at the location of the shock to the throat area.

Solution:

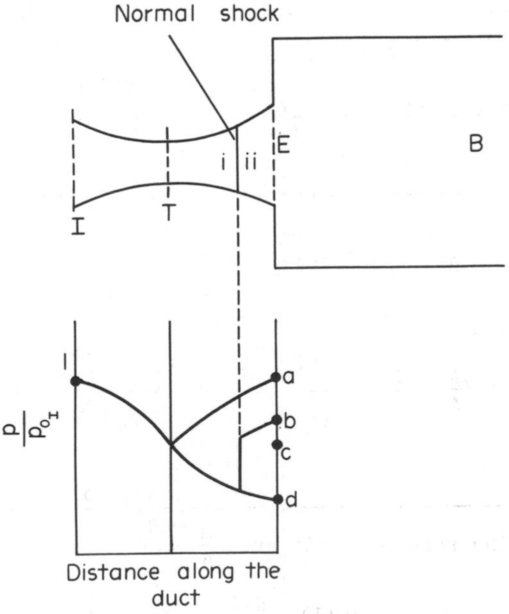

Fig. 3.5.

For isentropic choked flow

$$\left(\frac{\dot{m}\sqrt{c_pT_o}}{A\,p_o}\right)_E = \left(\frac{\dot{m}\sqrt{c_pT_o}}{A\,p_o}\right)_T \times \left(\frac{A_T}{A_E}\right)$$

From the isentropic flow table (table 2)

$$\left(\frac{\dot{m}\sqrt{c_pT_o}}{A\,p_o}\right)_T = 1.281 \;\therefore\; \left(\frac{\dot{m}\sqrt{c_pT_o}}{A\,p_o}\right)_E = 0.64$$

Therefore for subsonic exit (point a, Fig. 4.5)

$$\frac{p_E}{p_o} = 0.937,\; p_E = 0.937 \times 140 = 131\ \text{kN/m}^2,\ \text{too high}$$

For supersonic exit (point d)

$$\frac{p_E}{p_o} = 0.0938 \qquad p_E = 13.12\ \text{kN/m}^2,\ \text{too low}$$

Therefore there is a normal shock between the throat and the exit plane.

$$\frac{\dot{m}\sqrt{c_pT_o}}{A_Ep_E} = \frac{\dot{m}\sqrt{c_pT_o}}{A_T\,p_o} \times \frac{A_T}{A_E} \times \frac{p_o}{p_E} = 1.281 \times \tfrac{1}{2} \times 1.4 = 0.896$$

From the tables $M_E \simeq 0.4$ and $\dfrac{p_E}{p_{oE}} = 0.896$.

(Note that $\dfrac{\dot{m}\sqrt{c_pT_o}}{A\,p}$ is not tabulated but can be found by dividing $\dfrac{\dot{m}\sqrt{c_pT_o}}{A\,p_o}$ by $\dfrac{p}{p_o}$)

$$\therefore\; p_{oE} = \frac{100}{0.896} = 111.5\ \text{kN/m}^2.$$

Total pressure ratio across the shock $= \dfrac{111.5}{140} = 0.796.$

From tables for normal shocks

$M_i = 1.84,\; M_{ii} = 0.608$

From the isentropic flow tables

$$\frac{\dot{m}\sqrt{c_pT_o}}{A_i\,p_o} = 0.863.$$

$$\therefore\; \frac{A_i}{A_T} = \left(\frac{\dot{m}\sqrt{c_pT_o}}{A\,p_o}\right)_T \Big/ \left(\frac{\dot{m}\sqrt{c_pT_o}}{A\,p_o}\right)_i = \frac{1.281}{0.863} = 1.49.$$

3.4 Supersonic diffusers

It may be noted (section 2.1) that when flow at the entrance to a convergent-divergent passage is supersonic, then the fluid will be slowed down and compressed to a Mach number $M_T \geqslant 1$ at the throat. The convergent duct is then referred to as a supersonic diffuser. If the Mach number at the throat is unity the fluid may expand again to supersonic velocities, or be compressed beyond the throat at subsonic velocities. Supersonic diffusers are used in wind tunnels with a supersonic working section and in jet engine intakes. We now consider the flow in these devices and for simplicity ignore all losses except those occuring in normal shocks.

3.4.1 Wind tunnel with a supersonic diffuser and second throat

A simple configuration for the tunnel is shown in Fig. 3.6. It consists of a convergent-divergent duct, a working section, another convergent-divergent duct and a compressor. When the flow is established in the tunnel, the second throat could be matched to the first throat to give nearly isentropic flow throughout the tunnel sections. (This means throats of equal area in the absence of boundary layers). However during starting the second throat must be larger than the first throat as can be seen from the following considerations:

It is usual for the inlet stagnation pressure and temperature to be kept constant during starting while the back pressure (p_E) is gradually lowered. If the second throat is larger than the first throat, then the first throat chokes first and as the back pressure is lowered further a shock moves down the divergent portion of the inlet duct (section 3.2 above). However if flow can be choked at the second throat, then flow upstream of this throat cannot be affected any more by lowering the back pressure. If M reaches unity at the second throat then $\left(\frac{\dot{m}\sqrt{c_p T_o}}{A\, p_o} \right)$ must have the same value there as at the first throat. Since $\dot{m}$ and T_o are constant, and with a shock in the inlet duct or working section, p_o at the first throat is greater than p_o at the second throat, the second throat must be larger than the first to avoid choked flow at the second throat. The greatest reduction in total pressure occurs when the shock is standing in the test section, because the shock then occurs at the maximum possible Mach number and consequently produces the largest loss in total pressure. Therefore, neglecting the effect of the boundary layers, the area of the second throat A_{T_2} must be $\geqslant A_{T_1} \times \frac{p_{o1}}{p_{o2}}$ for

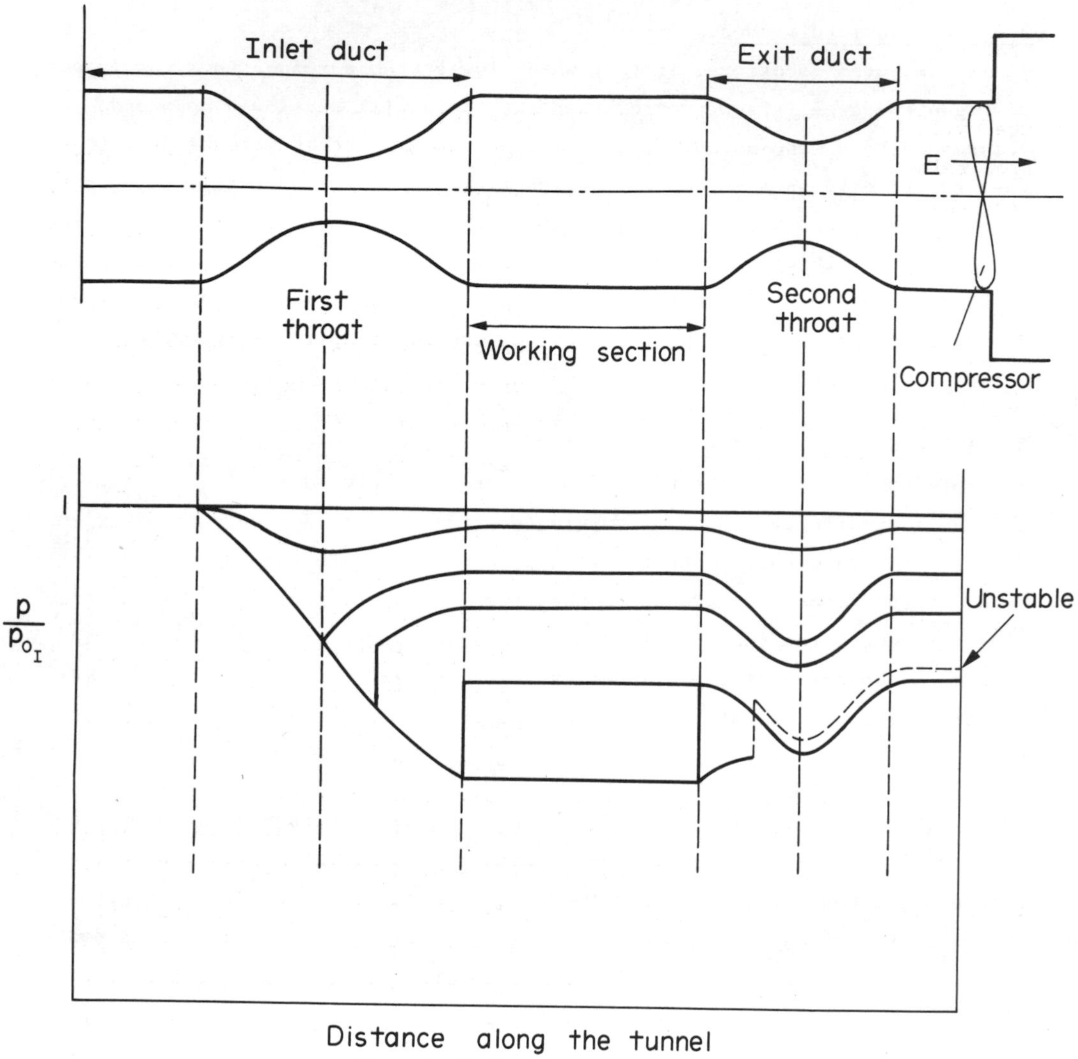

Fig. 3.6 Supersonic wind tunnel.

starting the tunnel, where $\frac{p_{o1}}{p_{o2}}$ is the total pressure ratio across the shock wave in the test section. (Note that p_{o2}/p_{o1} is the minimum pressure ratio that will start the tunnel.) If the pressure ratio $\frac{p_{o2}}{p_{o1}}$ is lowered slightly from this value, then the shock jumps from the working section to a position past the second throat where the area is approximately equal to the area of the working section. This is because a normal shock cannot remain stably in any position in a convergent passage (see section 3.5 below). The tunnel is now said to have started. The power required to run the tunnel is dependent on the loss of stagnation pressure across the shock wave. For most efficient working the shock should be at the diffuser throat because this is the point of minimum supersonic Mach number (and hence minimum $p_{o1} - p_{o2}$) downstream of the working section. This can be arranged if the area of the second throat can be varied.

3.4.2 Jet engine intakes

Supersonic diffusers are also used for the inlet of jet engines operating at supersonic speeds. These engines usually reach their operating speed by being accelerated from lower speeds (as distinct from overshooting the operating speed) and, as for the supersonic wind tunnel diffuser, operation of the inlet during starting must be investigated.

Consider a convergent-divergent inlet having an inlet area A_I and throat area A_T (Fig. 3.7i). When the inlet travels at supersonic speeds with no detached shock ahead of it, (Fig. 3.7ii) all the free stream flow corresponding to the cross-sectional area A_I enters the engine. (In supersonic flow the inlet cannot affect the flow upstream (section 1.9) and the streamlines are parallel up to the tip of the inlet). If the inlet cannot pass this flow, then a detached shock stands ahead of the inlet. The flow behind the shock is subsonic (in the vicinity of area A_I) and may spill over the tip of the inlet. To understand the formation of the detached shock, consider the case of variable throat area. If the duct is of constant area the flow proceeds with unchanged velocity. (Note that the present discussion assumes the back pressure to be sufficiently low so as not to affect flow in the duct). As the throat is made smaller, the Mach number at the throat decreases until at the critical point when $A_T = A^*$ and $M = 1$. Now if the area is further reduced the flow will be choked and the mass flow rate through the duct will be reduced. The reduction of the mass flow is caused by a shock wave which forms near the throat and travels upstream. The shock can be shown to be unstable in the convergent channel (see section 3.5 below) and it is almost immediately disgorged. It will stabilize upstream in front of the diffuser (Fig. 3.7iii). The portion of the shock facing the inlet is close to a normal shock and the flow after this part of the shock is subsonic (although flow may be supersonic behind the more oblique portions). Flow behind the portion of the shock facing the inlet first diverges and then converges to give M = 1 at the throat. (Note that the bending of the streamlines can only occur for a shock oblique to the flow). The divergence of the streamlines behind the shock causes some of the air in the path of the duct to be spilled around its lips, thus reducing the mass flow rate. If the throat area is further reduced, the shock moves further upstream and causes more spilling of the flow. Now if the above procedure is reversed, that is the throat is expanded again, there will be a hysteresis effect. This is caused by the shock which is now present near the mouth of the inlet.

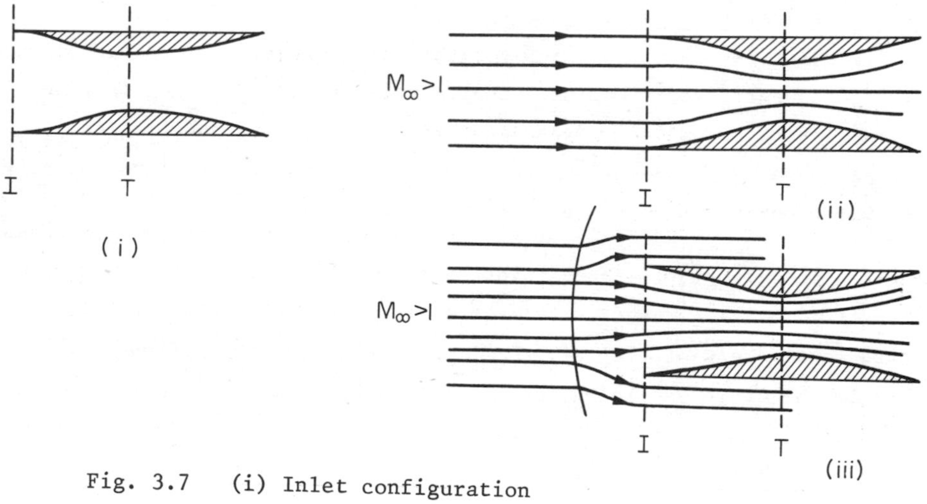

Fig. 3.7 (i) Inlet configuration

(ii) No detached shock, with all the free stream flow corresponding to A_I entering the inlet.

(iii) Detached shock ahead of the inlet, causing spill-over.

There is a loss of stagnation pressure through the shock and therefore the throat must be made larger for the shock to be swallowed than it was for the shock to form. When the shock is swallowed it stabilizes in the divergent part of the inlet and M is then greater than 1 at the throat. The throat area can now be reduced to move the shock nearer to the throat. This can be continued until the farthest stable point is reached, where M = 1 at the throat, the shock is weakest, and the efficiency is highest.

For a fixed geometry diffuser the hysteresis may be overcome by overspeeding. That is, the shock is swallowed by accelerating the diffuser beyond its design speed, and is then moved to its desired position by slowing down to the design speed.

A quantitative description of the hysteresis effect may be obtained from Fig. (3.8) which is similar to that given in reference 5. The lower curve is for isentropic flow with M = 1 at the throat and is plotted directly for the isentropic flow relations. The upper curve shows the maximum contraction possible with a shock standing at the mouth of the inlet, forcing all the air through the inlet, but with a reduced stagnation pressure.

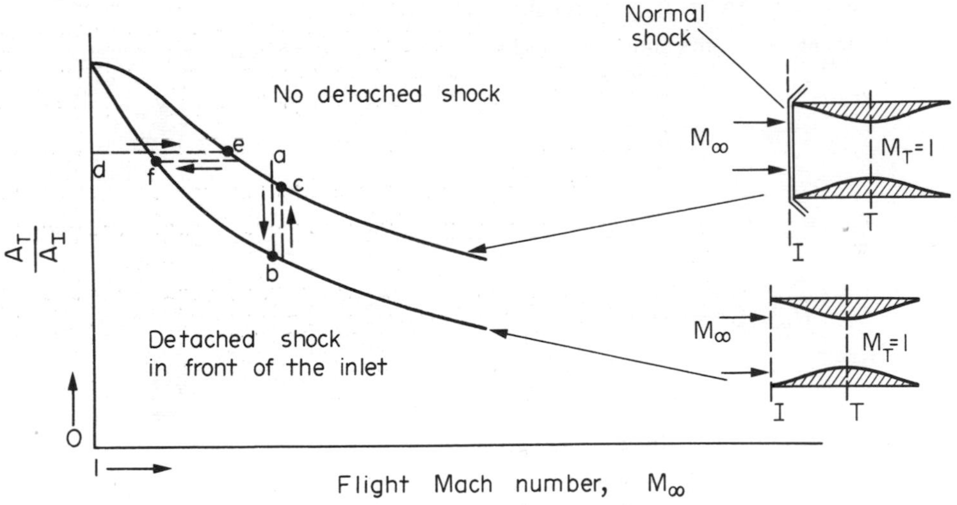

Fig. 3.8 Limiting contraction ratios versus flight Mach number.

The above discussion of the variable area diffuser can be represented by the lines parallel to the area ratio axis. As the throat area is decreased the state point moves down the line ab; At point b the shock is disgorged and stabilizes in front of the inlet. The increase in throat area then moves the state point up the line bc and at point c the shock is swallowed. Similarly the overspeeding of the constant geometry inlet can be represented by lines de and ef drawn parallel to the Mach number axis. It can be seen that a one-dimensional inlet accelerating towards its design speed (point f) starts off with a shock in front of it as $M = 1$ is passed. The shock gets closer to the mouth of the inlet as the design speed is approached. However for the shock to be swallowed, the intake should be accelerated beyond the design speed to point e where the shock is swallowed. The inlet can now be slowed down to the design speed.

3.5 Instability of a normal shock in a convergent passage

Consider a convergent-divergent duct placed in a supersonic stream (Fig. 3.9). If the throat area is sufficiently large then flow in the duct will be supersonic. However, as the throat is made smaller the Mach number there decreases until at the critical point when $A_T = A^*$, $M_T = 1$. If a further reduction in the throat area is attempted the flow is choked and the mass flow rate must therefore decrease. The reduction in the mass flow rate is by means of a shock wave which forms at the throat and travels upstream. The shock wave will be unstable in the convergent passage, because to maintain constant

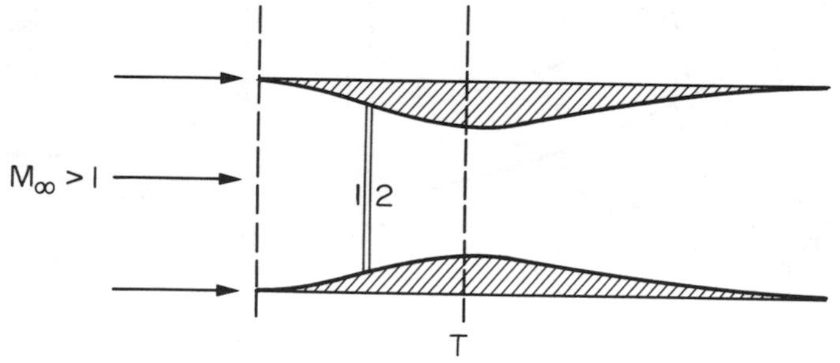

Fig. 3.9 Convergent-divergent duct in supersonic stream

mass flow rate and stagnation temperature across the shock, we must have (from equation 3.1)

$$p_{o_1} A_1^* = p_{o_2} A_2^* \tag{3.3}$$

However $A_2^* < A_1^*$ (throat area reduced from A_1^* to A_2^*) and $p_{o_2} < p_{o_1}$ across the shock, so that equation (3.3) cannot be satisfied if the shock stabilizes in the convergent passage; the shock is therefore immediately disgorged.

EXERCISES

1. The pressure p and density ρ of the burning gas in the convergent-divergent nozzle of a rocket satisfy the approximate equation $p/\rho = C$ (constant). The gas accelerates throughout the nozzle. At entry the gas is at low velocity and pressure p_o.

Fing the pressure at the throat. If the velocity at the exit is twice that at the throat, find the ratio (exit area)/(throat area). Neglect fluid friction.

Answer: $0.61\ p_o$, 2.24.

2. Air is supplied to a convergent-divergent nozzle from a reservoir where the pressure is 100 kN/m^2. The air is then discharged through a short pipe into another reservoir where the pressure can be varied. The cross-sectional area of the pipe is twice the area of the throat of the nozzle. Friction and heat transfer may be neglected throughout the flow.

If the discharge pipe has constant cross-sectional area, determine the range of static pressure in the pipe for which a normal shock will stand in the divergence of the nozzle.

If the discharge pipe tapers, so that its cross-sectional area is reduced by 25% along its length, show that a normal shock cannot be drawn to the end of the divergence of the nozzle. Find the maximum strength of shock (as expressed by the upstream Mach number) which can be formed.

Answer: 93.7 to 51.3 kN/m^2, 2.116.

3. A perfect gas, for which the ratio of the specific heat capacities is 1.4, flows through a convergent-divergent nozzle from a reservoir in which the stagnation pressure is p_o. The throat area of the nozzle is A^* and the area at the exit plane is $1.8\ A^*$.

A pitot tube mounted on the axis of the nozzle measures a pressure of $0.74\ p_o$ at the exit plane. The pitot tube is very small in relation to the size of the nozzle and can be moved upstream to measure the pitot pressure at a series of points along the axis.

Assuming that the flow is isentropic except across a shock wave, show that the recorded pitot pressure at the exit is not consistent with an assumption that the flow is supersonic throughout the divergent part of the nozzle, with a bow shock in front of the pitot tube. Find the area, in terms of A*, at the position in the nozzle at which there is a standing shock wave.

Make a sketch of the distribution of Mach number, stagnation pressure and pitot pressure along the nozzle. Determine the pitot pressure at the point downstream of the throat at which the area is 1.4 A*.

Answer: 1.63 A*, $0.829p_o$.

4. A convergent-divergent nozzle with a throat area of half the exit area is supplied with gas (γ = 1.4) at a stagnation pressure of 140 kN/m^2. It discharges into an atmosphere of static pressure 100 kN/m^2. Find the stagnation pressure of the gas at the exit from the nozzle.

Answer: 111.5 kN/m^2.

5. A supersonic wind tunnel has a working section of 2.5 m x 2.5 m. If the Mach number in the working section is to be 2.2, find the area of the first throat. By neglecting all losses except those due to essential normal shocks, estimate the size of the second throat and the power required to start the tunnel. Assume that the stagnation temperature and pressure at the inlet are 290 K and 100 kN/m^2, and that the geometry of the tunnel is not varied while the tunnel is running.

Answer: 3.115 m^2, 4.96 m^2, 30700 kW.

6. A fixed convergent-divergent engine intake is designed so that at maximum air flow a normal shock stands just upstream of the mouth in flight at a Mach number of 1.5. What is the ratio of the mouth area to throat area? Neglecting friction, find the maximum possible mass flow rate into the engine, for 0.1 (m^2) throat area, in flight at an altitude of 10,000 meters at M = 1.5, just before the shock is swallowed, and at a slightly higher speed with the shock swallowed.

Answer: 1.093, 20.3 kg/s.

7. The air intake of a jet engine consists of a convergent-divergent passage of fixed geometry.

Find the maximum possible mass flow rate into the engine, for unit throat area, in terms of ground-level atmospheric pressure p_g and density ρ_g, for the following duties:

(i) at rest on the ground,

(ii) in flight at a Mach number of 0.4 near ground level,

(iii) in flight at an altitude where the pressure is $\frac{1}{4}p_g$ and the density $\frac{1}{3}\rho_g$, and at a Mach number of 1.6, assuming that the shock is then just about to be swallowed,

(iv) under virtually the same conditions as (iii) but at a slightly higher speed so that the shock is swallowed.

The effects of friction and heat transfer are to be neglected. For the air assume $\gamma = 1.4$.

Answers: $0.685\sqrt{p_g\rho_g}$, $0.752\sqrt{p_g\rho_g}$, $0.611\sqrt{p_g\rho_g}$, $0.611\sqrt{p_g\rho_g}$.

8. An aircraft is flying at a Mach number of 2 and at a height of 19,000 m. The intake of an engine swallows an amount of air equal to that which would flow through an area of 0.2 m^2 at the flight Mach number with the static conditions ahead of the aircraft. In the first part of the intake there is a 5% loss in stagnation pressure (due to inclined shock waves) after which the flow is still supersonic. The subsequent flow in the intake passage can be treated as one-dimensional, frictionless and adiabatic.

If the operating condition of the engine is such that the value of $(m\sqrt{T_o})/p_o$ at entry will be 0.0052 ms $\sqrt{K}$, show that there must be a normal shock wave in the intake passage and determine the Mach number immediately upstream of it.

For stability of operation, the passage is designed so that the cross-sectional area is at a minimum upstream of the normal shock wave, and this minimum is 5% less than the area at the shock wave. Find the Mach number and the area at the minimum cross-section.

Answers: 1.355, 1.222, 0.13 m^2.

CHAPTER 4

FLOW WITH FRICTION OR HEAT TRANSFER

4.1 Introduction

In the previous chapters, the flow was assumed to be adiabatic and the analysis was based on the assumption of isentropic flow with appropriate procedures to account for irreversibility across shock waves. However, in cases where friction or heat transfer is the major factor bringing about the changes in fluid properties, appropriate treatment is required. A simple analysis can be carried out for flow with heat transfer if it is assumed that frictional effects are negligible (see 4.3 for the implications of this assumption). The flow with friction is irreversible, but the effects of the irreversibility can be taken into account from a knowledge of the behaviour of friction. In both cases the flow is assumed to be one-dimensional.

4.2 Adiabatic flow in a duct of constant cross-sectional area with friction

4.2.1 The Fanno curves.

From the continuity equation (1.2)

$$G = \frac{\dot{m}}{A} = \rho V = \text{constant} \tag{4.1}$$

and from the energy equation (1.11)

$$h_o = h + \frac{V^2}{2} = \text{constant} \tag{4.2}$$

Note that these equations are valid for both reversible and irreversible flows. In the case of flow with friction the equations are valid whether friction is present or not.

From (4.1) and (4.2)

$$h = h_o - \frac{G^2}{2\rho^2} \tag{4.3}$$

For a pure substance, the entropy, s, can be expressed as a function of two properties e.g.

$$s = f(h, \rho) \tag{4.4}$$

From equations (4.3) and (4.4)

$s = f(h, G/\sqrt{2(h_o - h)})$

Therefore a curve of h against s can be plotted for the flow process. This is called a Fanno curve. Figs. (4.1) and (4.2) illustrate families of Fanno curves. (Note that both s and h are usually given relative to the values at a datum state. However a change in the datum would only change the origin and will not change the shape of the curves).

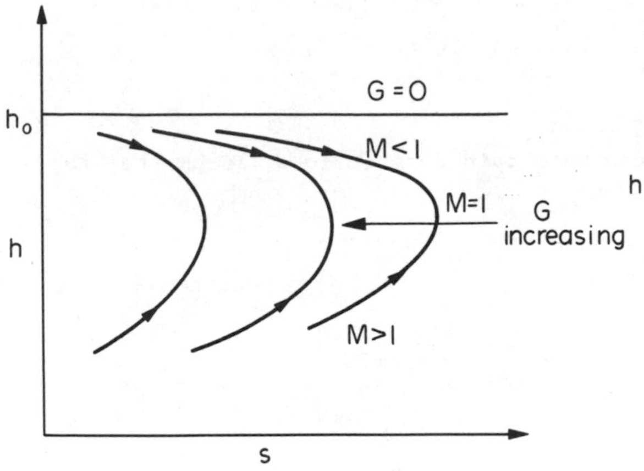

Fig. 4.1 Fanno curves with the same stagnation enthalpy but different values of G.

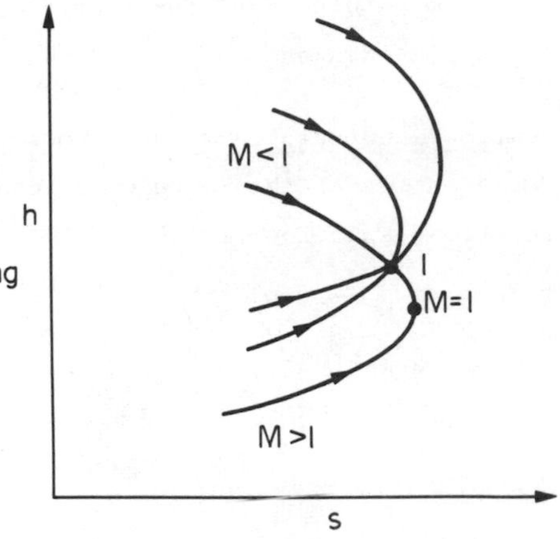

Fig. 4.2 Fanno curves with the same h_1 and s_1 but different G and h_o.

The entropy of a fluid element must increase as it travels along the duct if there is adiabatic flow with friction. Therefore the path of the state of the fluid element must be towards the right along any of the Fanno curves. It will now be shown that the Mach number is equal to 1 at the point of maximum entropy on a Fanno curve.

From the Gibb's relation, equation (1.14)

$$T\,ds = dh - \frac{dp}{\rho},$$

and from (4.3)

$$dh = G^2 \frac{d\rho}{\rho^3}$$

so that $T\,ds = G^2 \dfrac{d\rho}{\rho^3} - \dfrac{dp}{\rho}$

At the point of maximum entropy on a Fanno curve, ds = 0 and the above equation gives:

$$\left(\frac{\partial p}{\partial \rho}\right)_s = \frac{G^2}{\rho^2} = V^2$$

However $\left(\frac{\partial p}{\partial \rho}\right)_s = a^2$, therefore M = 1 at the point of maximum entropy.

Frictional effects in adiabatic flow in a constant area duct cause the state point to move continuously along a Fanno curve towards M = 1. The process will now be discussed for a perfect gas, for which analytical expressions for the state changes along the Fanno curves can be derived.

4.2.2 Fanno relations for a perfect gas

The equation of the Fanno curve can be obtained as follows: From the Gibb's relation, equation (1.14)

$$T\,ds = dh - \frac{dp}{\rho},$$

and for a perfect gas

$$p/\rho = RT \text{ i.e., } \frac{dp}{p} = \frac{d\rho}{\rho} + \frac{dT}{T},$$

and $dh = c_p\, dT$.

$$\text{Therefore } \frac{ds}{R} = \frac{dh}{RT} - \frac{dp}{\rho RT} = \frac{dh}{RT} - \frac{dp}{p} = \frac{dh}{RT} - \frac{dT}{T} - \frac{d\rho}{\rho} = \frac{dh}{RT} - \frac{dh}{c_p T} - \frac{d\rho}{\rho}$$

From the continuity equation (4.1), $\frac{d\rho}{\rho} + \frac{dV}{V} = 0$, therefore the above equation can be written as

$$\frac{ds}{R} = \frac{dh}{RT} - \frac{dh}{c_p T} + \frac{dV}{V}$$

From the energy equation (4.2) $dh + V\,dV = 0$, therefore

$$\frac{dV}{V} = -\frac{dh}{V^2} = -\frac{dh}{2(h_o - h)},$$

also

$$c_p = \frac{\gamma R}{\gamma - 1}$$

so that

$$\frac{ds}{R} = \frac{1}{\gamma}\frac{dh}{RT} - \frac{dh}{2(h_o - h)}. \tag{4.5a}$$

Now

$$\gamma RT = a^2 = a_o^2 - (\gamma - 1)(h_o - h)$$

therefore

$$\frac{ds}{R} = \frac{dh}{a_o^2 - (\gamma-1)(h_o - h)} - \frac{dh}{2(h_o - h)}$$

and

$$\frac{s - s_1}{R} = \ln \left\{ \frac{\left[a_o^2 - (\gamma-1)(h_o - h)\right]^{\frac{1}{\gamma-1}} (h_o - h)^{\frac{1}{2}}}{\left[a_o^2 - (\gamma-1)(h_o - h_1)\right]^{\frac{1}{\gamma-1}} (h_o - h_1)^{\frac{1}{2}}} \right\} \qquad (4.5)$$

This equation can be used to plot a Fanno curve for given conditions at station 1. Noting that $h_o - h = c_p(T_o - T)$ for a perfect gas, equation (4.5) becomes

$$\frac{s - s_1}{R} = \ln\left\{ \left(\frac{T}{T_1}\right)^{\frac{1}{\gamma-1}} \left[\frac{T_o - T}{T_o - T_1}\right]^{\frac{1}{2}} \right\}$$

This equation can be used to plot a Fanno curve for temperature against entropy as shown in Fig. (4.3). For example, if ρ_1, V_1 and T_1 are specified, then,

$$a_1 = \sqrt{\gamma R T_1}\ , \quad M_1 = \frac{V_1}{a_1} \quad \text{and} \quad T_o = T_1\left(1 + \frac{\gamma-1}{2} M_1^2\right).$$

The value of s_1 relative to its value at the datum can be found from ρ_1 and T_1. A curve of s against T can be plotted for the specified $G = \rho_1 V_1$.

The curve can also be plotted by using the equations derived below, as follows: For a specified M_1 the temperature ratio $\frac{T}{T_1}$ can be found from (4.6), and $(s - s_1)$ from (4.10), for various values of M. For given T_1, $s - s_1$ can be plotted against T.

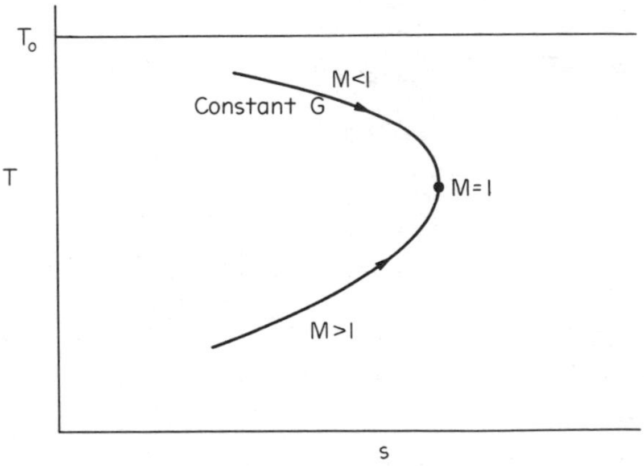

Fig. 4.3 Temperature-entropy diagram for a perfect gas.

The maximum value of entropy is reached at $M = 1$, therefore the flow will always proceed in the direction shown on the curve i.e. towards the point at which $M = 1$. The slope of the Fanno curve can be found from equation (4.5a) i.e.

$$\frac{\partial s}{\partial h} = R\left(\frac{1}{a^2} - \frac{1}{V^2}\right) = R\left(\frac{M^2 - 1}{V^2}\right)$$

$$\frac{\partial T}{\partial s} = \frac{1}{c_p}\left(\frac{\partial h}{\partial s}\right) = \frac{\gamma}{c_p}\,\frac{M^2 T}{M^2 - 1}$$

This equation shows that at $M = 1$, entropy is a maximum ($(\frac{\partial T}{\partial s}) = \infty$) as already shown in section 4.2.1. For the lower part of the curve in Fig. 4.3, $\frac{\partial T}{\partial s} > 0$, i.e. $M > 1$ and similarly for the upper part of the curve $M < 1$. Friction will accelerate a subsonic flow with decreasing total pressure and will decelerate a supersonic flow with decreasing total pressure. In both cases the flow tends towards Mach number unity.

The change in the flow properties can be expressed in terms of the change in the Mach number as follows:

Temperature

$$dT_o = 0, \quad T\left(1 + \frac{\gamma-1}{2}M^2\right) = \text{constant} \tag{4.6a}$$

Therefore

$$\frac{T}{T_1} = \frac{1 + \frac{\gamma-1}{2}M_1^2}{1 + \frac{\gamma-1}{2}M^2} \tag{4.6}$$

Pressure

$$G = \rho V = \frac{p}{RT}Ma = \frac{p}{RT}M\sqrt{\gamma RT} = \text{constant}$$

Therefore

$$\frac{pM}{\sqrt{T}} = \text{constant} \tag{4.7a}$$

or $\frac{p}{p_1} = \frac{M_1}{M}\sqrt{\frac{T}{T_1}}$.

Considering (4.6)

$$\frac{p}{p_1} = \frac{M_1}{M}\sqrt{\frac{1 + \frac{\gamma-1}{2}M_1^2}{1 + \frac{\gamma-1}{2}M^2}} \tag{4.7}$$

Density

$$\frac{\rho}{\rho_1} = \frac{p}{p_1} \frac{T_1}{T}$$

Considering (4.6) and (4.7)

$$\frac{\rho}{\rho_1} = \frac{M_1}{M} \left[\frac{1 + \frac{\gamma-1}{2} M^2}{1 + \frac{\gamma-1}{2} M_1^{\,2}} \right]^{\frac{1}{2}} \tag{4.8}$$

Entropy Change

From (1.14)

$$\frac{ds}{R} = \frac{dh}{RT} - \frac{dp}{p}$$

For a perfect gas $dh = c_p \; dT = \frac{\gamma R}{\gamma-1} dT$. Therefore,

$$\frac{ds}{R} = \frac{\gamma}{\gamma-1} \frac{dT}{T} - \frac{dp}{p} \; .$$

From (4.6a)

$$\frac{dT}{T} = \frac{- \, d \left(1 + \frac{\gamma-1}{2} M^2\right)}{\left(1 + \frac{\gamma-1}{2} M^2\right)} \tag{4.9a}$$

and from (4.7a)

$$\frac{dp}{p} = \frac{1}{2} \frac{dT}{T} - \frac{dM}{M} = - \frac{1}{2} \frac{d \left(1 + \frac{\gamma-1}{2} M^2\right)}{\left(1 + \frac{\gamma-1}{2} M^2\right)} - \frac{dM}{M} \tag{4.9b}$$

Therefore

$$\frac{ds}{R} = \frac{-(\gamma+1)}{2(\gamma-1)} \frac{d \left(1 + \frac{\gamma-1}{2} M^2\right)}{\left(1 + \frac{\gamma-1}{2} M^2\right)} + \frac{dM}{M}$$

Integration of this expression gives,

$$\frac{s - s_1}{R} = - \ln \left[\frac{M_1}{M} \left(\frac{1 + \frac{\gamma-1}{2} M^2}{1 + \frac{\gamma-1}{2} M_1^{\,2}} \right) \right]^{\frac{\gamma+1}{2(\gamma-1)}} \tag{4.10}$$

Stagnation pressure. From the above analysis

$$ds = c_p \frac{dT}{T} - R \frac{dp}{p} = R \left(\frac{\gamma}{\gamma-1} \frac{dT}{T} - \frac{dp}{p} \right)$$

Also, by definition $\frac{p_o}{p} = \left(\frac{T_o}{T} \right)^{\frac{\gamma}{\gamma-1}}$

or

$$\frac{dp_o}{p_o} - \frac{dp}{p} = \frac{\gamma}{\gamma-1}\left(\frac{dT_o}{T_o} - \frac{dT}{T}\right) = -\frac{\gamma}{\gamma-1}\frac{dT}{T}$$

Therefore

$$ds = -R\frac{dp_o}{p_o}$$

and

$$\frac{p_o}{p_{o_1}} = e^{-(s - s_1)/R} \tag{4.11}$$

The Impulse function I

$$I = pA + \rho AV^2 = pA(1 + \frac{\rho V^2}{p}) = pA(1 + \frac{V^2}{RT}) = pA(1 + \gamma M^2)$$

Therefore

$$\frac{I}{I_1} = (\frac{p}{p_1})(\frac{1 + \gamma M^2}{1 + \gamma M_1^2}) = \frac{M_1}{M}(\frac{1 + \gamma M^2}{1 + \gamma M_1^2})\left[\frac{1 + \frac{\gamma-1}{2}M_1^2}{1 + \frac{\gamma-1}{2}M^2}\right]^{\frac{1}{2}} \tag{4.12}$$

4.2.3 Reference state and Fanno tables

It is convenient to calculate the above dimensionless ratios for various Mach numbers for use in solving numerical problems and for future reference (Table 3) as was done for isentropic flow. A convenient reference state for this purpose is where M = 1; an asterisk is used to denote this state. Note that this is a different state from that used in the isentropic flow tables where the condition M = 1 is reached with the entropy kept constant. The variations of the ratios with Mach number are also shown in Fig. (4.4).

4.2.4 Choking in flow with friction

In sections 4.2.1 and 4.2.2 it was shown that, for flow with friction, the Mach number increases along the pipe for subsonic flow and decreases for supersonic flow. In both cases the maximum value of entropy is reached at M = 1. Therefor there is a maximum length of the pipe for which the Mach number of unity is reached at the outlet of the pipe for specified inlet conditions. The flow is then said to be choked. If the duct length is increased from this maximum value, then the flow within the duct must be adjusted to keep M = 1 at the outlet from the pipe. In the case of subsonic flow, the adjustment is achieved by means of an automatic reduction in the

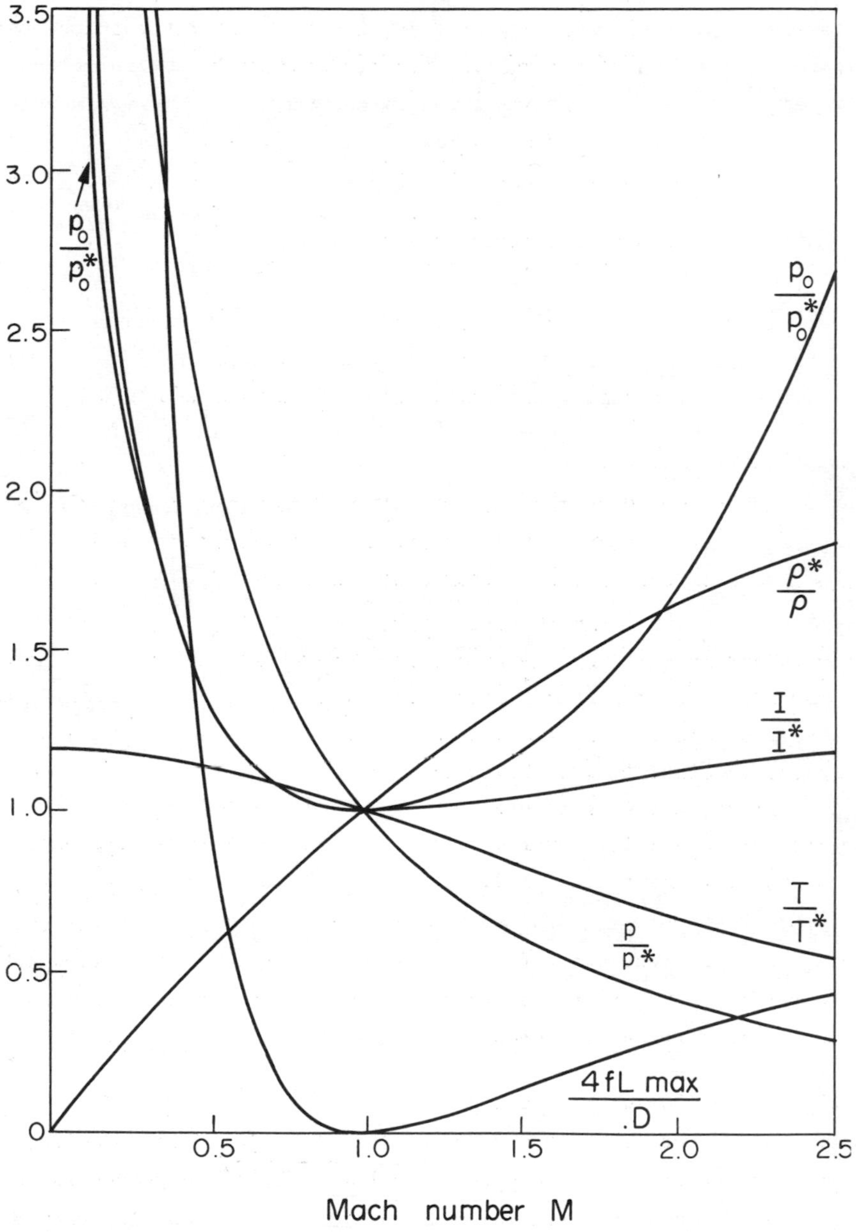

Fig. 4.4 Adiabatic flow in a duct of constant cross-sectional area with friction - Mach number function ($\gamma = 1.4$).

mass flow rate just sufficient to give a Mach number of unity at the outlet. For supersonic flow the adjustment is usually accompanied by formation of shock waves. Further details are given in reference 5. The maximum length of the pipe can be calculated as follows:

The momentum equation (1.23) gives

$$\frac{dp}{p} + \frac{\gamma}{2}\,dM^2 + \frac{\gamma M^2}{2}\frac{dT}{T} + \frac{\gamma M^2}{2}\,4f\,\frac{dx}{D} = 0.$$

Substituting for $\frac{dp}{p}$ and $\frac{dT}{T}$ from (4.9a) and (4.9b), gives

$$4f\,\frac{dx}{D} = \frac{\gamma+1}{2\gamma}\,\frac{d(1+\frac{\gamma-1}{2}M^2)}{1+\frac{\gamma-1}{2}M^2} + \frac{2dM}{\gamma M^3} - \frac{\gamma+1}{2\gamma}\,\frac{dM^2}{M^2}.$$

Assuming the friction factor f ($= \tau_w/(\frac{1}{2}\rho V^2)$) to be a constant, this equation can be integrated to give,

$$4f\,\frac{x - x_1}{D} = \frac{1}{\gamma}\left(\frac{1}{M_1^2} - \frac{1}{M^2}\right) + \frac{\gamma+1}{2\gamma}\ln\left\{\frac{M_1^2}{M^2}\left[\left(\frac{1+\frac{\gamma-1}{2}M^2}{1+\frac{\gamma-1}{2}M_1^2}\right)\right]\right\} \qquad (4.13)$$

If the variation in f along the duct is significant then an average value of the friction factor, $\bar{f}$, can be defined as

$$\frac{\bar{f}(x - x_1)}{D} = \int_{x_1}^{x} f\,\frac{dx}{D}$$

and f in equation 4.13 can then be replaced by $\bar{f}$. Experimental evidence[8] shows that the friction factor varies with the Reynolds number R_e ($= \frac{\rho VD}{\mu}$) in fully developed pipe flow as shown in Fig. 4.5.

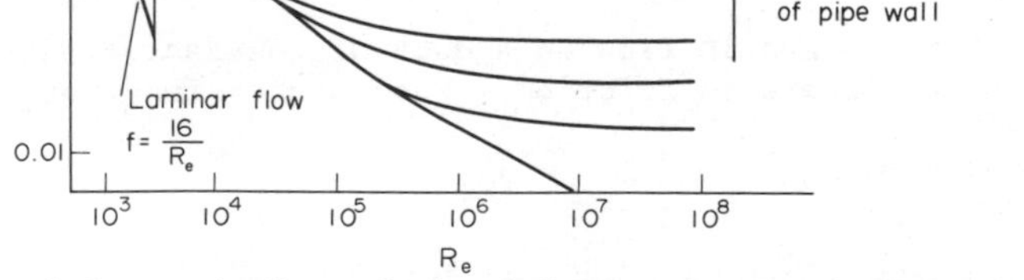

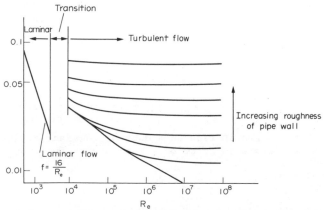

Fig. 4.5 Variation of the friction factor f with the Reynolds number R_e.

Thus, for fully developed turbulent flow at large Reynolds number, f can be considered to be approximately constant. This is a good assumption in many practical cases.

The effect of compressibility on the local value of f has been discussed in reference 9. Experimental results show that the compressibility effect is not significant for a fully developed, subsonic turbulent flow (i.e., distance from the duct inlet > 50 pipe diameters). However, for supersonic flow the local value of f is likely to be significantly dependent on Mach number, Reynolds number, and development of the boundary layers etc.

The length of the pipe reaches a maximum (L_{max}), for given conditions at station 1, and M = 1 at the exit since no entropy increase is possible beyond M = 1. Substituting M = 1 in equation (4.13) gives

$$4f \frac{x^* - x_1}{D} = 4f \frac{L_{max}}{D} = \frac{1}{\gamma} \left(\frac{1}{M_1^2} - 1 \right) + \frac{\gamma+1}{2\gamma} \ln \left(\frac{\frac{\gamma+1}{2} M_1^2}{1 + \frac{\gamma-1}{2} M_1^2} \right)$$

Fig. (4.6) illustrates the variation of $\frac{L_{max}}{D}$ with M_1

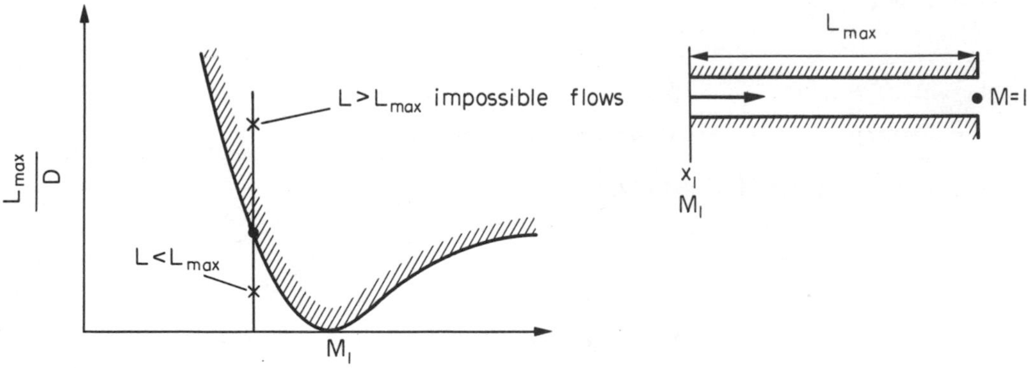

Fig. 4.6.

<u>4.3 Flow in a duct of constant cross-sectional area with heat transfer</u>

In this case the analysis is based on the assumption that effects of viscosity are negligible. This idealised process will be referred to as simple heating or simple cooling. The heat transfer causes a change in the stagnation temperature of the flowing fluid.

The idealised analysis cannot always be justified for flows encountered in practice. Frictional effects will be present because of the connection between mechanisms of fluid friction and heat transfer, if heat is transferred by forced convection from the walls of the duct (see e.g. reference 12, the Reynolds analogy). General formulae for the numerical integration for flows with friction and heat transfer are given in reference 11. If the change in the stagnation temperature is caused by combusion of either a flowing combustible mixture, or a fuel discharged into the flowing fluid, then, in addition to viscous effects, there will be a change in the chemical composition of the fluid. Clearly, the results can be expected to have a high degree of validity if departures from the assumptions of the model are small. For example, in the case of combustion, if the duct is relatively short and the change in stagnation temperature is large, then effects of viscosity on the flow may be neglected.

4.3.1 Fundamental equations for simple heating or simple cooling - The Rayleigh flow

The steady flow momentum equation (1.4) gives

$$dp + \rho V dV = 0$$

This may be combined with the continuity equation

$$\frac{\dot{m}}{A} = \rho V = G$$

to give

$$dp + G\ dV = 0.$$

For constant G, this equation can be integrated to give

$$p + GV = p + \rho V^2 = \frac{I}{A} = \text{constant}$$

or

$$p + \frac{G^2}{\rho} = \text{constant} \tag{4.14}$$

The above equations simply state that the impulse function $A(p + \rho V^2)$ is constant for this flow as may be expected from (2.14).

For a pure substance both p and ρ can be expressed as functions of the enthalpy h and the entropy s so that, for given constants G and $p + \rho V^2$, equation (4.14) may be used to plot the variations of h with s. The resulting curve is known as a "Rayleigh curve" (Fig. 4.7).

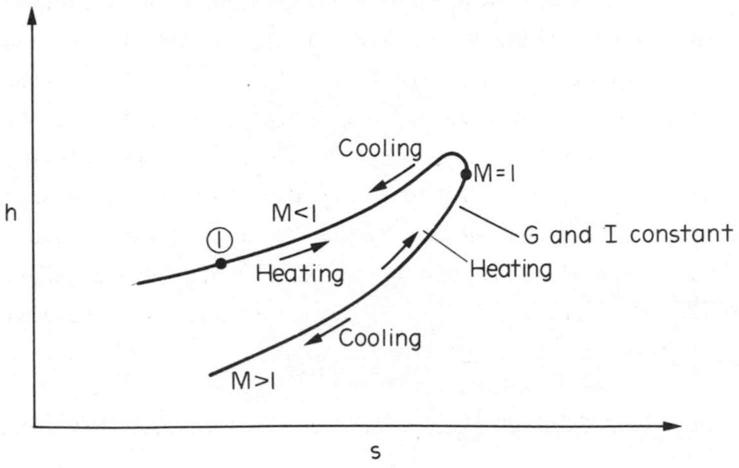

Fig. 4.7 The Rayleigh Curve

The idealised process is thermodynamically reversible i.e., $đQ = Tds$. Therefore heat addition causes an entropy increase for either subsonic or supersonic flow. Note the similarity in this case to flow with friction. Conversely heat rejection causes an entropy decrease. Now from (4.14)

$$dp - G^2 \frac{d\rho}{\rho^2} = 0$$

or $\frac{dp}{d\rho} = V^2$

At the point of maximum entropy, $ds = 0$. Since p can be expressed as a function of ρ and s, then

$$\frac{dp}{d\rho} = \left(\frac{\partial p}{\partial \rho}\right)_s + \left(\frac{\partial p}{\partial s}\right)_\rho \frac{ds}{d\rho} = \left(\frac{\partial p}{\partial \rho}\right)_s = a^2 = V^2$$

i.e., $M = 1$ at the point of maximum entropy. Therefore the flow cannot continue beyond $M = 1$ with either heat addition or heat rejection alone. However the combination of heating and cooling can allow a flow to cross $M = 1$ without discontinuity. Note the similarity to the combination of area convergence and divergence in isentropic flow. Also for given inlet conditions there is a maximum value of heat input corresponding to $M = 1$ at the outlet. If the amount of heat added is greater than this value, the flow will be choked.

That is the inlet Mach number will be reduced in magnitude to a value consistent with the specified amount of heat input. Choking will be discussed further in 4.3.5. Equation 4.14 shows a linear variation of p with $\frac{1}{\rho} = v$, having a negative slope (Fig. 4.8). The line is usually called the Rayleigh

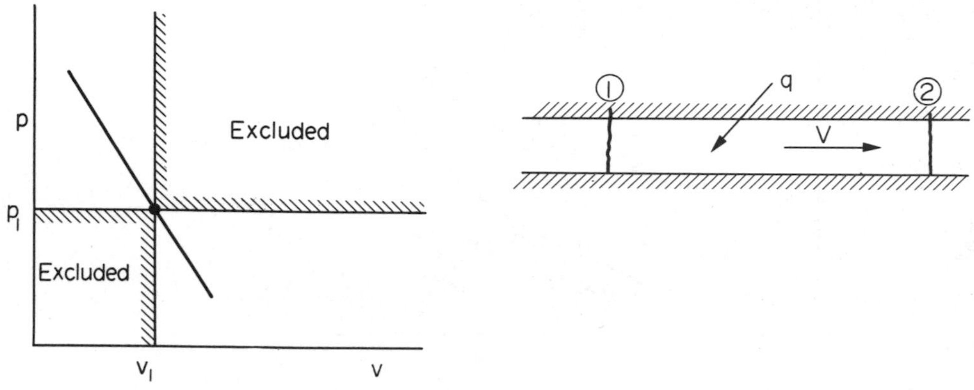

Fig. 4.8 The Rayleigh Line

line in the English literature and the Mikhel'son line in the Russian literature. There are two restricted areas in Fig. 4.8 corresponding to a positive slope of the line.

4.3.2 The Rayleigh flow for a perfect gas

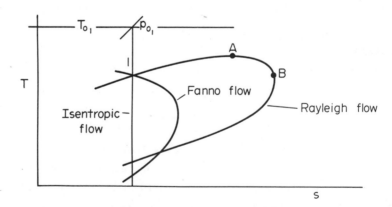

Fig. 4.9 T - s diagram for Rayleigh flow of a perfect gas

The Rayleigh curve for a perfect gas is shown in Fig. 4.9. The curve can be plotted by using equations (4.20) and (4.23). Values of T_2 and s_2 can be

computed from these equations for given conditions at station 1 and a chosen value of M_2. The slope of the curve can be obtained as follows:
From equation (1.14)

$$ds = c_p \frac{dT}{T} - R\frac{dp}{p}$$

$$\frac{ds}{dT} = \frac{c_p}{T} - \frac{R}{p}\left(\frac{dp}{dT}\right)$$

From (4.14)

$$p + \rho V^2 = p(1 + \gamma M^2) = \text{constant}$$

Therefore

$$\frac{dp}{p} = \frac{-d(1 + \gamma M^2)}{1 + \gamma M^2} = \frac{-2\gamma M dM}{1 + \gamma M^2}$$

The continuity equation can be written as

$$\rho V = \frac{p}{RT} M\sqrt{\gamma RT} = \text{constant}$$

$$\text{or } \frac{dT}{T} = 2\left(\frac{dp}{p} + \frac{dM}{M}\right) = 2\left(\frac{-2\gamma M dM}{1 + \gamma M^2} + \frac{dM}{M}\right)$$

$$= 2\left(\frac{dM - \gamma M^2 dM}{M(1 + \gamma M^2)}\right)$$

Therefore

$$\left(\frac{\partial p}{\partial T}\right)_{\text{Rayleigh}} = -\frac{\gamma M^2 p}{(1 - \gamma M^2)T}$$

and

$$\left(\frac{\partial s}{\partial T}\right)_{\text{Rayleigh}} = \frac{c_p}{T} + \frac{R}{p}\left(\frac{\gamma M^2 p}{(1 - \gamma M^2)T}\right)$$

$$= \frac{c_p}{T}\left(1 + \frac{\gamma - 1}{\gamma}\,\frac{\gamma M^2}{1 - \gamma M^2}\right)$$

$$= \frac{c_p}{T}\left(\frac{1 - M^2}{1 - \gamma M^2}\right) \qquad (4.15)$$

Equation (4.15) shows that at $M = 1$, $\left(\frac{\partial T}{\partial s}\right)_{\text{Rayleigh}} = \infty$. Also at $M = \frac{1}{\sqrt{\gamma}}$ point, A in Fig. 4.9, the slope is zero. In the upper part of the curve, M is always

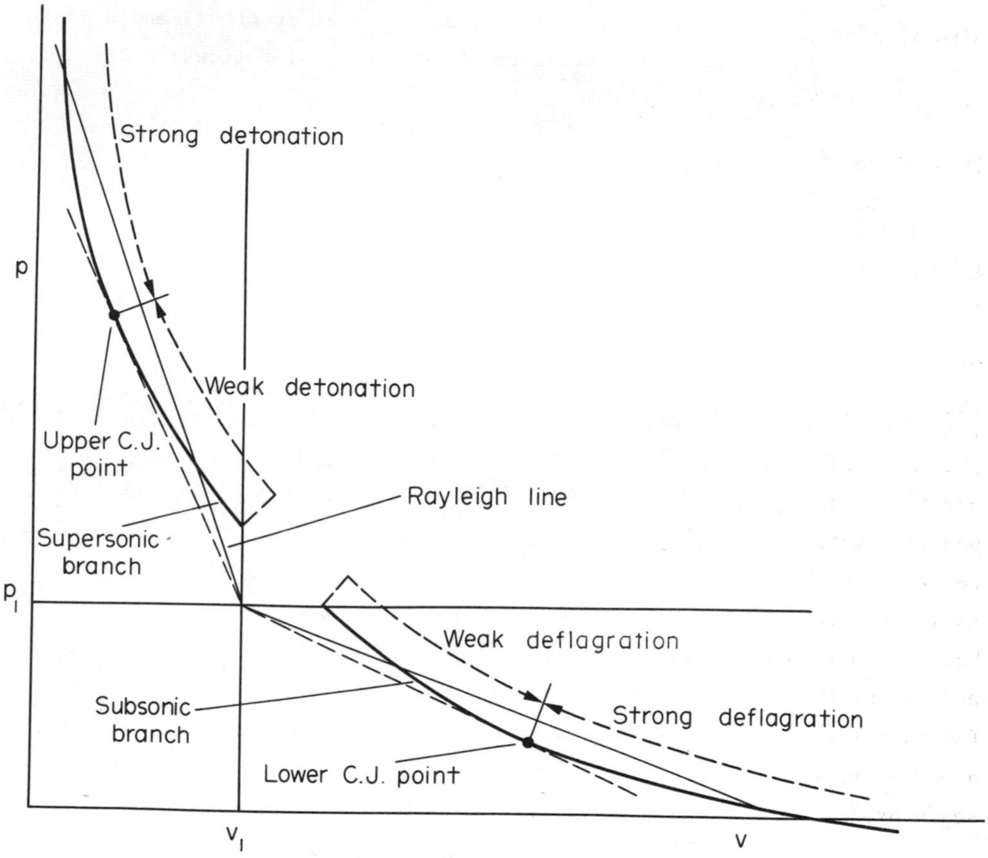

Fig. 4.10 p-v diagram for frictionless flow in a duct of constant cross-sectional area with heat addition.

< 1, because; (a) $\frac{\partial T}{\partial s} > 0$ for $M < \frac{1}{\sqrt{\gamma}}$ up to point A, and (b) $\frac{\partial T}{\partial s} < 0$ between point A, where $M = \frac{1}{\sqrt{\gamma}}$ and point B where M = 1. In the lower part of the curve $\frac{\partial T}{\partial s} > 0$ so that M > 1.

The process of simple heating can be represented on a p-v diagram for a perfect gas as follows:

From the steady flow energy equation (1.11)

$$h_2 + \frac{V_2^2}{2} = h_1 + \frac{V_1^2}{2} + q \tag{4.16}$$

where q = heat transfer per unit mass, between stations 1 and 2 (Fig. 4.8).

Also from equation (4.14)

$$(\rho_1 V_1)^2 = (\rho_2 V_2)^2 = \frac{p_2 - p_1}{v_1 - v_2} \tag{4.17}$$

This gives

$$V_1^2 - V_2^2 = (p_2 - p_1)(v_1 + v_2)$$

substituting in equation (4.16) gives

$$h_2 - h_1 - q = \tfrac{1}{2}(p_2 - p_1)(v_1 + v_2)$$

or

$$\frac{\gamma}{\gamma-1}(p_2 v_2 - p_1 v_1) - q = \tfrac{1}{2}(p_2 - p_1)(v_1 + v_2) \tag{4.18}$$

The locus of final states (p_2, v_2) for an initial state (p_1, v_1) and heat addition q can be determined from (4.18). The state 2 is obtained by the point of intersection of the Rayleigh line (equation 4.17) and the graph of equation (4.18). Referring to Fig. (4.10) it can be seen that with heat addition (finite q), in the velocity region above a subsonic velocity and below a supersonic velocity, there are no solutions to the equations. Above and below these two critical velocities there are two possible end states for each inlet velocity. However, due to the intersection of the curve with $p = 0$ axis, very low subsonic velocities yield only one solution on the subsonic branch.

The entire process is referred to as deflagration if $M_1 < 1$ and detonation if $M_1 > 1$. Fig. 4.10 illustrates that for deflagration there is a pressure drop between stations 1 and 2 ($p_2 < p_1$). The magnitude of the pressure drop is greater for strong deflagration than for weak deflagration. In the case of detonation, the pressure rises between stations 1 and 2 and the pressure rise is greater for strong detonation than for weak detonation (see references 9 and 10 for a more detailed description).

The justification for the physical existence of these solutions cannot be made wholly on the basis of an examination of the equations. The stability of any of the above solutions can be determined by investigating the true transient behaviour in the neighbourhood of these steady solutions.

It can be shown that the point of tangency of the curve and the Rayleigh line (Chapman - Jouguet (C.J) points) corresponding to sonic flow at station 2.

4.3.3 Simple heating relations for a perfect gas

(duct of constant cross-sectional area, no friction)

The flow properties may be expressed as functions of the Mach numbers, as was done for isentropic flow and flow with friction, as follows:

Pressure From (4.14)

$$p + \rho V^2 = p(1 + \gamma M^2) = \text{constant}$$

Therefore
$$\frac{p_2}{p_1} = \frac{1 + \gamma M_1^2}{1 + \gamma M_2^2} \tag{4.19}$$

Temperature

The continuity equation can be written as

$$\rho V = \frac{pM}{RT}\sqrt{\gamma RT} = \text{constant}$$

Therefore, $\frac{dp}{p} + \frac{dM}{M} - \frac{1}{2}\frac{dT}{T} = 0.$

or
$$\frac{T_2}{T_1} = \left(\frac{p_2 M_2}{p_1 M_1}\right)^2 = \frac{M_2^2}{M_1^2}\left(\frac{1 + \gamma M_1^2}{1 + \gamma M_2^2}\right)^2 \tag{4.20}$$

Density

$$\frac{\rho_2}{\rho_1} = \frac{V_1}{V_2} = \left(\frac{p_2}{p_1}\right)\left(\frac{T_1}{T_2}\right) = \frac{M_1^2}{M_2^2}\left(\frac{1 + \gamma M_2^2}{1 + \gamma M_1^2}\right)$$

Stagnation Temperature

By definition of the stagnation state,

$$T_o = T\left(1 + \frac{\gamma-1}{2}M^2\right)$$

Therefore
$$\frac{T_{o_2}}{T_{o_1}} = \frac{M_2^2(1 + \gamma M_1^2)^2 \left(1 + \frac{\gamma-1}{2}M_2^2\right)}{M_1^2(1 + \gamma M_2^2)^2 \left(1 + \frac{\gamma-1}{2}M_1^2\right)} \tag{4.21}$$

Stagnation Pressure

By definition

$$p_o = p\left(\frac{T_o}{T}\right)^{\frac{\gamma}{\gamma-1}}$$

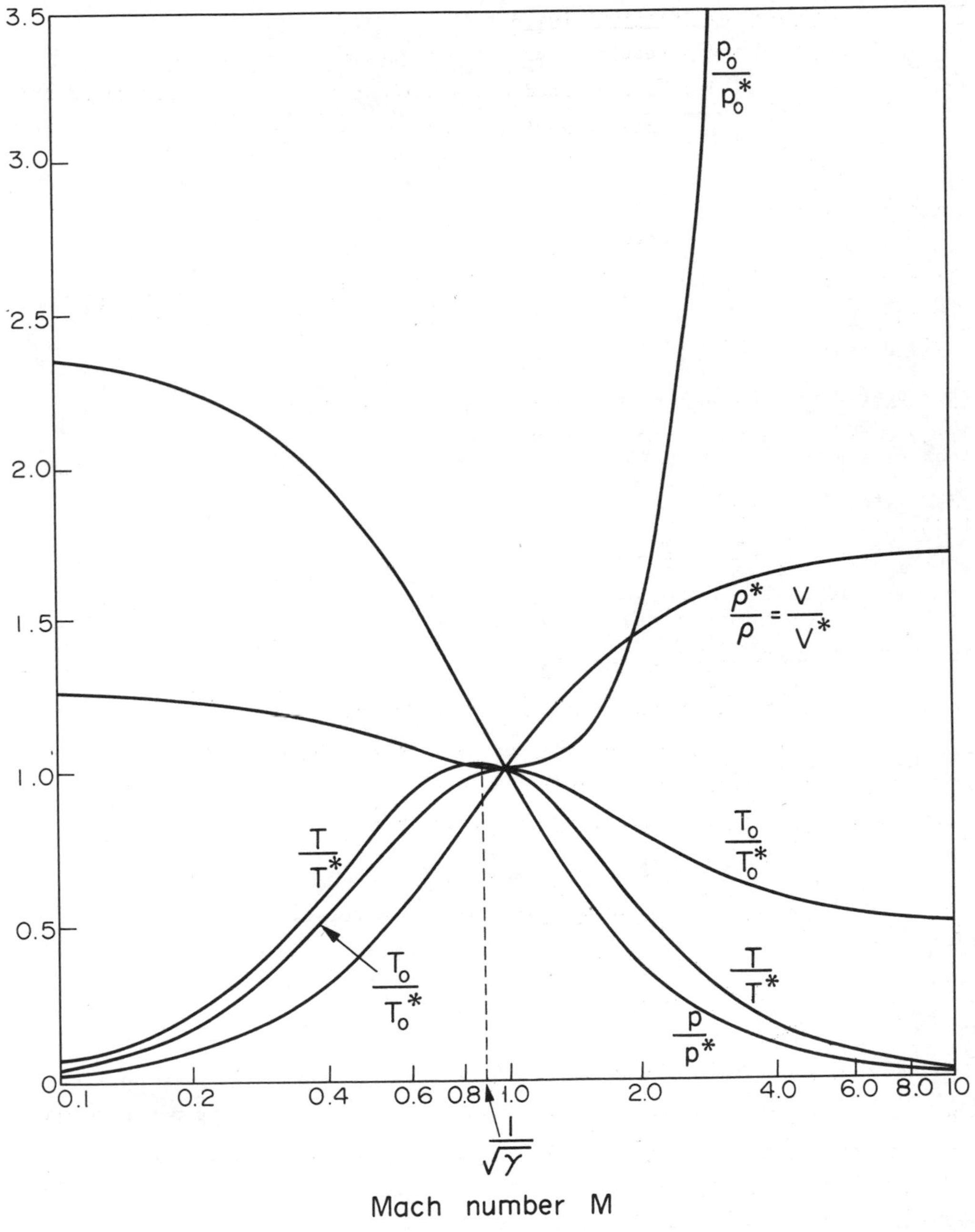

Fig. 4.11 Frictionless flow in a duct of constant cross-sectional area with heat transfer. Mach number functions (γ = 1.4).

Therefore from (4.19), (4.20) and (4.21),

$$\frac{p_{o_2}}{p_{o_1}} = \left(\frac{1+\gamma M_1^{\,2}}{1+\gamma M_2^{\,2}}\right)\left[\frac{1+\frac{\gamma-1}{2}M_2^{\,2}}{1+\frac{\gamma-1}{2}M_1^{\,2}}\right]^{\frac{\gamma}{\gamma-1}} \tag{4.22}$$

Entropy change

From (1.14)

$$Tds = dh - \frac{dp}{\rho}$$

For a perfect gas

$$dh = c_p dT = \frac{\gamma R}{\gamma-1}\,dT \text{ and } \rho = \frac{p}{RT}$$

Therefore $ds = \dfrac{\gamma R}{\gamma-1}\dfrac{dT}{T} - R\dfrac{dp}{p}$

Integration gives

$$\frac{s_2 - s_1}{R} = \ln\left(\frac{T_2}{T_1}\right)^{\frac{\gamma}{\gamma-1}} - \ln\frac{p_2}{p_1}$$

Using (4.19) and (4.10)

$$\frac{s_2 - s_1}{R} = \ln\left(\frac{M_2}{M_1}\right)^{\frac{2\gamma}{\gamma-1}}\left[\frac{1+\gamma M_1^{\,2}}{1+\gamma M_2^{\,2}}\right]^{\left(\frac{\gamma+1}{\gamma-1}\right)} \tag{4.23}$$

4.3.4 Reference state and Rayleigh tables

The point of maximum entropy (M = 1) is chosen as reference and properties there are denoted by an asterisk. The non-dimensional quantities given by equations (4.19) to (4.23) can be expressed in terms of Mach number (e.g. $\frac{p}{p^*} = \frac{\gamma+1}{1+\gamma M^2}$). These are tabulated for given Mach numbers in Table 4. The variations are also shown in Fig. (4.11). It is interesting to note (Fig. 4.11) that in a small region from $M = \frac{1}{\sqrt{\gamma}}$ to M = 1, the static temperature actually decreases as heat is added to accelerate the flow towards M = 1.

4.3.5 Choking in flow with heat transfer

It was noted in 4.3.1 that there is a maximum amount of heat that can be added to a flow and that in the limiting case, the exit Mach number will be exactly unity. If this amount of heat is exceeded, the flow in the pipe would be adjusted to give M = 1 at the exit and the flow is then said to be choked. In the case of supersonic flow, for a given heat addition and given inlet conditions, there is a maximum allowable Mach number at the inlet, for which steady flow (with no shock waves) is possible.

The choking effects must be considered for combustion chambers of ramjets, jet engines, etc., since the chamber entrance velocities are limited for a given heat addition. Also Fig. 4.11 shows that for combustion chambers operating near choking stagnation pressure drops are large. These pressure drops do not include frictional losses. They are due to losses which may be caused by heat addition.

Note that heat removed in the supersonic case (in the absence of shock discontinuities) has a maximum value corresponding to a given inlet Mach number since the leaving flow cannot have a temperature below zero (corresponding to leaving $M = \infty$).

4.3.6 Explosion waves in combustible mixtures

The results of the simple heating analysis can be used to study "steady explosion waves". The flow on both sides of such waves is steady relative to an observer moving with the wave front. Some quantities of interest are; (a) speed of the wave relative to the unburned gas, (b) the motion of the burned gas, and (c) the change in pressure and temperature across the wave.

Example:

It is found experimentally that when a detonation wave is propagated into a combustible gas mixture at rest, then the velocity of flow behind the wave is just sonic, relative to the wave front. Show that of all possible donation waves, this is the wave which propagates at the lowest speed. Assume that the gases have the properties of a single perfect gas, and that the rise of relative stagnation temperature in the wave is constant.

If the gases have initially a temperature of $5^{o}C$ and a pressure of 100 kN/m^2, and are assumed to have the same properties as air, and the rise in relative stagnation temperature is $1110^{o}C$, what is the static pressure and the static temperature behind such a wave?

Consider the flow relative to the wave. Since I, $\dot{m}$ and c_p are constant

$$\left(\frac{I}{\dot{m}\sqrt{c_pT_o}}\right)_2 = \left(\frac{I}{\dot{m}\sqrt{c_pT_o}}\right)_1 \times \sqrt{\frac{T_{o_1}}{T_{o_2}}}$$

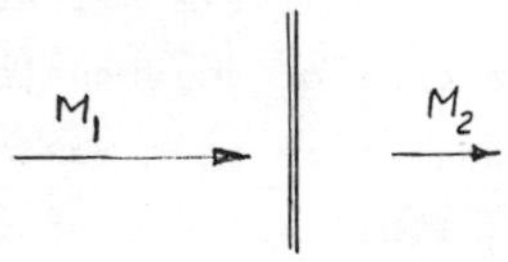

detonation wave

let $T_{o_2} = T_{o_1} + \Delta T_o$

Expressing $\frac{I}{\dot{m}\sqrt{c_pT_o}}$ in terms of M (table 2.3) gives

$$\frac{1}{M_2}(1+\gamma M_2^2)\left\{1+\frac{\gamma-1}{2}M_2^2\right\}^{-\frac{1}{2}} = \frac{1}{M_1}(1+\gamma M_1^2)\left\{1+\frac{\gamma-1}{2}M_1^2\right\}^{-\frac{1}{2}}\left(1+\frac{\Delta T_o}{T_{o_1}}\right)^{-\frac{1}{2}}$$

Noting that $T_o = T(1 + \frac{\gamma-1}{2}M^2)$, the above equation gives,

$$\frac{M_2^2}{(1+\gamma M_2^2)^2}\left(1+\frac{\gamma-1}{2}M_2^2\right) = \frac{M_1^2}{(1+\gamma M_1^2)^2}\left(1+\frac{\gamma-1}{2}M_1^2+\frac{\Delta T_o}{T_1}\right) \qquad (4.24)$$

This relationship (for given T_o and T_1) is of the form shown below.

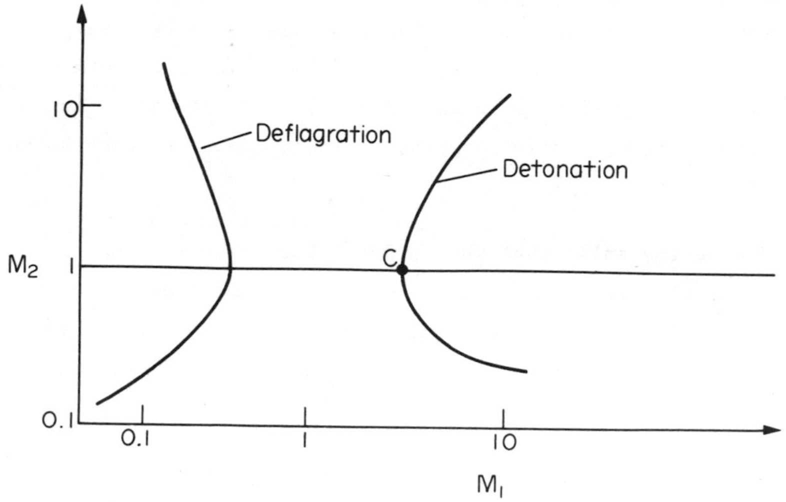

We have to show that

$\frac{dM_1}{dM_2} = 0$ at $M_2 = 1$ (point C on the diagram).

Differentiating equation 4.24 logarithmically

$$\left(\frac{1}{M_2^2} - \frac{2\gamma}{1 + \gamma M_2^2} + \frac{(\frac{\gamma-1}{2})}{1 + \frac{\gamma-1}{2} M_2^2} \right)$$

$$= \left(\frac{1}{M_1^2} - \frac{2\gamma}{1 + \gamma M_1^2} + \frac{(\frac{\gamma-1}{2})}{1 + \frac{\gamma-1}{2} M_1^2} \right) \frac{d(M_1^2)}{d(M_2^2)} = 0.$$

Therefore

$$(1 + \gamma M_2^2)(1 + \frac{\gamma-1}{2} M_2^2) - 2\gamma M_2^2(1 + \frac{\gamma-1}{2} M_2^2) + \frac{\gamma-1}{2} M_2^2(1 + \gamma M_2^2) = 0.$$

This gives $1 - M_2^2 = 0$ or $\underline{M_2 = 1}$

$T_1 = 5^\circ C = 278$ K $\quad \frac{\Delta T_o}{T_1} = \frac{1110}{278} \simeq 4.0 \quad \gamma = 1.4$

substitution in equation 4.24, noting that $M_2 = 1$, gives

$$M_1^4 - 21.2\, M_1^2 + 1 = 0$$

$$M_1^2 = 21.153 \text{ (or } 0.047)$$

$$M_1 = 4.6$$

$\left(\frac{T_1}{T_{o_1}} \right) = 0.192$ from table 2

$$T_{o_1} = \frac{278}{0.192} = 1450 \text{ K}$$

$$T_{o_2} = 1450 + 1110 = 2560 \text{ K}$$

$\left(\frac{T_2}{T_{o_2}} \right) = 0.833$ from table 2 ($M_2 = 1$)

$$T_2 = 1862^\circ C$$

$$p_2 = \frac{p_1 \left(\frac{\dot{m}\sqrt{c_p T_o}}{A\, p_o} \right)_1 \left(\frac{p}{p_{o_2}} \right)_2 \sqrt{\frac{T_{o_2}}{T_{o_1}}}}{\left(\frac{p}{p_o} \right)_1 \left(\frac{\dot{m}\sqrt{c_p T_o}}{A\, p_o} \right)_2}$$

Using table 2

$$p_2 = \frac{100 \times 23.3 \times 0.528}{1.281} \sqrt{\frac{2560}{1450}} = 1275 kN/m^2$$

EXERCISES

1. An insulated pipe of length L and diameter D discharges into a region where the static pressure is 400 kN/m^2. Air flows into the pipe from a convergent-divergent nozzle with throat area half that of the pipe. The stagnation pressure at the nozzle inlet is 1000 kN/m^2. The friction coefficient f for the pipe is 0.005, and L/D is 20.

Show that a normal shock stands in the pipe, and calculate the Mach number of the flow at exit from the pipe.

Answer: 0.69

2. Show that for laminar, incompressible, fully developed flow in a pipe of circular cross-section and diameter D the friction factor is given by
$f = (4\pi\mu/\dot{m})D$,
where μ is the viscosity and $\dot{m}$ the mass flow rate. It may be assumed without proof that the velocity profile is parabolic.

Air at an absolute pressure of 500 kN/m^2 and temperature of 300 K, contained in a pressure vessel which has a wall thickness of 6 mm, is found to be leaking at a rate of 2×10^{-6} kg/s to the atmosphere. Assuming that there is a small circular hole of length equal to the wall thickness, determine its diameter. The above expression for the friction factor may be used (entry effects and the influence of density variation on the expression for f being neglected) with the viscosity constant and equal to 3×10^{-5} kg/m s. It may be assumed that the flow is choked at outlet from the hole, and that acceleration of the flow up to the entrance of the hole is isentropic.

Answer: 0.065 mm.

3. (a) Estimate the maximum flow rate of air through the passage shown, assuming that the friction coefficient of the duct is 0.005.

(b) For what range of back pressures will this maximum flow rate be achieved?

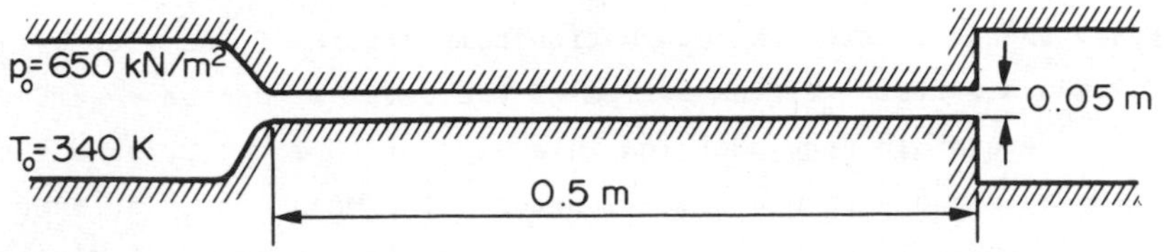

Answer: 2.56 kg/s; $p \leqslant 315$ kN/m^2.

4. A perfect gas of specific heat ratio 2 flows through a duct of constant cross-section in which it is "heated" by an electric discharge. Wall friction, heat losses and magnetic forces are negligible. The gas enters with M = 0.707 and the electrical input is raised to the maximum permissible level. Find the ratio (increase in K.E.) / (electrical input).

Answer: 7:3.

5. Show that at constant mass flow rate, the heat addition model for steady one-dimensional flow (no friction or area changes) gives maximum entropy at both the lower and the upper Chapman- ouguet points.

6. Air flows without friction in a constant area duct. At the duct entrance M = 2, p = 138 kN/m^2 and T = 38°C. As the air moves along the duct, heat is removed as long as it is possible to do so.
(a) Calculate $\dot{m}/A$, the mass rate of flow per unit area.
(b) If no shock occurs what is the maximum velocity that is reached?
(c) If a shock does occur somewhere and the same amount of heat is removed as in (b), what is the final temperature?

Answer: (b) 834 m/s; (c) 60.5^{o}C.

7. The exhaust gases from an engine flow into a large tank and from there to atmosphere through a pipe of 40 mm bore. Conditions in the tank may be assumed steady, with the gases having a temperature of 200°C and negligible velocity. Friction and heat transfer in the pipe may be neglected, and at entry there is negligible loss of stagnation pressure.

If the gases may be treated as perfect, with c_p = 1.15 kJ/kg K and γ = 1.333, show that, for a flow rate of 0.1 kg/s and an exit pressure of 100 kN/m^2, the Mach number in the pipe is about 0.25, and determine the pressure in the tank.

The reduce pollution a catalyst is placed in the pipe, causing the unburnt fuel to react and so raising the stagnation temperature of the flow by 450 K between inlet and outlet. If the action of the catalyst can be regarded as equivalent to reversible heat addition in a pipe of 40 mm bore, show that, at the same flow rate and exit pressure as before, the Mach number at exit is raised to 0.35. Calculate the change of pressure in the tank.

Answer: 104 kN/m^2, increase of 8.4 kN/m^2.

8. A perfect gas ($\gamma = 1.4$, $c_p = 1.01$ kJ/kg K) enters a frictionless convergent-divergent nozzle at stagnation conditions of 300 kN/m^2 and 300 K. The gas then flows along a frictionless pipe of the same diameter as the nozzle outlet and exhausts into a large reservoir in which the pressure is adjustable. The area of the pipe is four times the throat area of the nozzle. Heat is added in the pipe to raise the stagnation temperature to 400 K. Find the mass flow rate per unit area of the pipe and the pressure in the reservoir for the cases where

(a) a weak normal shock is formed just downstream of the throat,

(b) a normal shock stands at the exit of the pipe.

Answer: (a) 174.5 kg/m^2s, 292.4 kN/m^2, (b) 174.5 kg/m^2s, 72.5 kN/m^2.

9. (a) Heat is being added to air as it flows frictionlessly through a pipe of varying cross-sectional area. If the pressure throughout the pipe is to be maintained constant, show that the relation between the area and the stagnation temperature is of the form

$$\frac{A}{A^*} = \frac{\gamma + 1}{2}\,\frac{T_o}{T_o^*} - \frac{\gamma - 1}{2}\;.$$

What is the meaning of A^* and T_o^*.

(b) Air enters a pipe 0.1 m diameter at a stagnation pressure and stagnation temperature of 100 kN/m^2 and 20^oC. The air flow is 0.3 kg/s and 300 kJ/s of heat is added at constant pressure. Neglecting friction, find the outlet diameter and the final Mach number.

Answer: 0.209 m, 0.045.

10. Air flows through a pipe of variable cross-sectional area in which the static pressure is kept constant throughout by heat transfer. If frictional effects may be neglected, show that the velocity is constant along the pipe,

and find an expression for the total rate of heat flow in terms of the static pressure, the velocity, the ratio of specific heats γ, and the change in area of the pipe.

At the entrance to the pipe, the diameter is 0.1 m and the stagnation pressure and temperature are 100 kN/m^2 and 20^oC. At the exit the Mach number is 0.2. If the rate of mass fow is 0.4 kg/s, find the rate of heat transfer and the diameter and stagnation pressure at the exit. For the air assume γ = 1.4 and c_p = 1.006 kJ/kg K.

Answer: $\frac{\gamma}{\gamma-1}\, pV(A_2 - A_1)$, -70.7 kJ/s, 0.0632 m, 101.7 kN/m^2.

11. A perfect gas flows at constant temperature along a duct of constant cross-sectional area with hydraulic mean diameter D and friction factor f($\tau_\omega/\frac{1}{2}\rho V^2$). Assuming one-dimensional flow, show that the pressure p, density ρ, velocity V and distance x are related by

$$\frac{dp}{p} = \frac{d\rho}{\rho} = -\frac{dV}{V} = -\frac{\gamma M^2}{2(1-\gamma M^2)}\left(4f\,\frac{dx}{D}\right).$$

Write down the limiting value of Mach number implied by these equations.

A perfect gas for which γ = 1.4 is pumped through a straight circular pipe, with internal diameter 1 m and friction factor 0.005, connecting two compressor stations 700 m apart. It may be assumed that there is sufficient heat transfer through the pipe wall to maintain the gas at 15^oC. At entry to the pipe, the stagnation pressure is 600 kN/m^2 and the Mach number is below the limiting value. Show that, in the limiting case of highest possible mass-flow rate, the Mach number at inlet is approximately 0.2, and determine the corresponding values of the flow rate and static pressure at outlet.

For the same inlet Mach number and stagnation pressure and the same friction factor, what would be the pressure at the pipe outlet if the flow were adiabatic?

Answer: $M = \frac{1}{\sqrt{\gamma}}$ at outlet, $\dot{m}$ = 376 kg/s, p = 138.5 kN/m^2, p = 189 kN/m^2.

12. The jet pipe on a jet-engine installation is 10 m long, 0.50 m in diameter, and terminates in a short converging nozzle with an exit diameter of 0.46 m. The stagnation temperature and pressure at entry to the pipe are

800°C and 300 kN/m^2 respectively. The engine is run with the aircraft stationary on the ground, and atmospheric pressure is 101.3 kN/m^2. Calculate the net thrust of the engine, (a) when skin friction in the jet pipe is neglected, and (b) when the skin friction coefficient ($\tau_\omega/\frac{1}{2}\rho V^2$) in the jet pipe is 0.01. The gas may be assumed to be a perfect gas with γ = 1.333 and c_p = 1.150 kJ/kg K.

Answer: 46 kN, 37.1 kN.

13. Air with a stagnation temperature of 288 K and a stagnation pressure of 101.3 kN/m^2 flows steadily through a short convergent nozzle and then through a duct with a uniform area of 0.1 m^2. It then exhausts to a large vessel in which the pressure is very low. In the first half of the duct heat is added at the rate of 1.0 MW, and in the second half of the duct heat is removed at the same rate. The flow is without friction. Calculate the mass-flow rate and the Mach number at exit from the duct.

Answer: 21.06 kg/s, 1.7.

CHAPTER 5

ONE-DIMENSIONAL UNSTEADY FLOW

5.1 Propagation of small disturbances

The propagation of a small disturbance (small wave) in an inviscid compressible fluid in a frictionless duct with no heat conduction will be considered (Fig. 5.1.A). It will be assumed that the cross-sectional area of the duct is constant and that the fluid ahead of the wave is moving with constant velocity (V) and has uniform properties p,ρ,T, etc. Also the changes in the fluid properties across the wave are assumed to be infinitesimal. It was shown in (1.8) that the speed of propagation of an infinitesimal disturbance in the fluid initially at rest is

$\left(\sqrt{\left(\frac{\partial p}{\partial \rho}\right)_s} = a\right)$ and that $a = \sqrt{\frac{\gamma p}{\rho}} = \sqrt{\gamma RT}$ for a perfect gas.

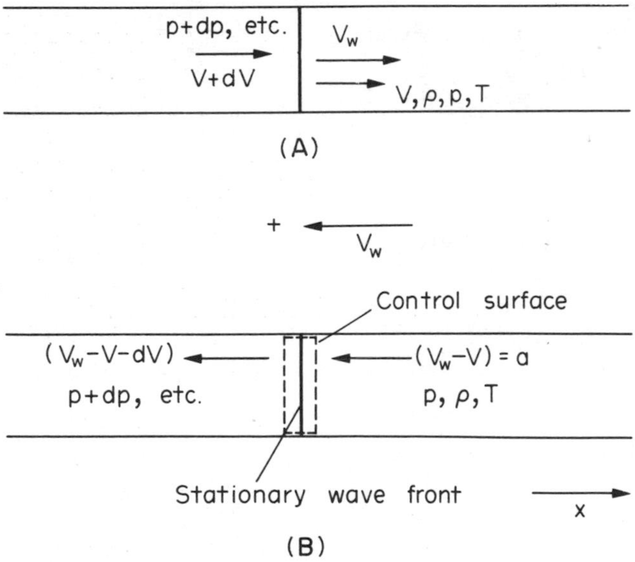

Fig. 5.1 Propagation of a small disturbance

If the fluid is initially moving with velocity V the wave can be brought to rest by superimposing a velocity of magnitude $V_w = V + a$ in the opposite direction (Fig. 5.1.B) and the same analysis can be carried out.

Therefore, in this case

$$V_w - V = a = \sqrt{\left(\frac{\partial p}{\partial \rho}\right)_s}$$

and from equations (1.16) and (1.17)

$$\frac{d\rho}{\rho} = \frac{dV}{a} \tag{5.1}$$

$$dV = \frac{dp}{\rho a} \tag{5.2}$$

(A similar analysis can be carried out for cylindrical and spherical disturbances, giving

$$a = \sqrt{\left(\frac{\partial p}{\partial \rho}\right)_s} \;)$$

The change of state due to small disturbances is isentropic (see 1.8) and for isentropic changes of a perfect gas

$$\frac{p^{\frac{\gamma-1}{\gamma}}}{T} = \text{constant or } \frac{p^{\frac{\gamma-1}{2\gamma}}}{a} = \text{constant}$$

$$\text{Therefore } \frac{dp}{p} = \frac{\gamma}{\gamma-1}\frac{dT}{T} = \frac{2\gamma}{\gamma-1}\frac{da}{a} \tag{5.3}$$

Substituting for dp in (5.2) gives

$$dV = \frac{dp}{\rho a} = \frac{2\gamma}{\gamma-1}\frac{da}{a}\left(\frac{p}{\rho a}\right) = \frac{2}{\gamma-1}\, da$$

Therefore

$$da = \frac{\gamma-1}{2}\, dV \tag{5.4}$$

across a wave moving with velocity

$$V_w = V + a \tag{5.5}$$

If the wave is propagating in the opposite direction to the flow (Fig. 5.2.A) then the wave can be brought to rest by superimposing a velocity in the opposite direction of magnitude $V_w = a - V$ (Fig. 5.2.B). Referring to (Fig. 5.2.B) the continuity equation gives

$$\rho a = (\rho + d\rho)(a + dV)$$

$$\text{or } \frac{d\rho}{\rho} = -\frac{dV}{a} \tag{5.6}$$

and the momentum equation gives

$$p - (p + dp) = \rho a(a + dV - a)$$

$$\text{or } dp = -\rho a\, dV \tag{5.7}$$

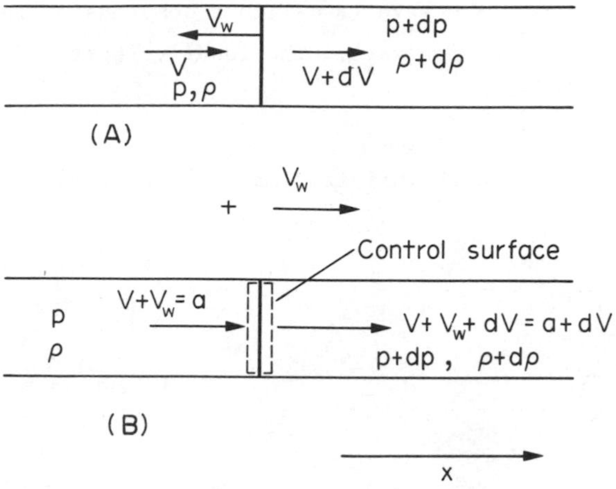

Fig. 5.2 Propagation of a small disturbance

Substituting for dp from (5.3) gives

$$da = - \frac{\gamma-1}{2} dV \tag{5.8}$$

across a wave propagating with velocity

$$V_w = V - a \tag{5.9}$$

(Note direction of increasing x on Fig. 5.2)

The above analysis applies to infinitesimal disturbances. However the resulting equations may be used to gain insight into the behaviour of small disturbances (i.e., $\frac{\delta p}{p} << 1$).

5.2 Unsteady flow with finite changes in fluid properties

Consider the propagation of disturbances (waves) in an inviscid compressible fluid in a frictionless duct of constant cross-sectional area with no heat conduction. The fluid properties (including the entropy) and velocity are considered to be uniform throughout the fluid before the passage of the waves.

5.2.1 Simple waves

In this case the waves propagate in one direction only (see Fig. 5.3). Consider a simple wave of finite amplitude, propagating along a duct of constant cross-sectional area. The wave can be considered to be made up of a series of infinitesimal waves, each propagating at the local velocity of sound relative to the fluid. If the pressure increases at a point x during the passage of the wave, the wave is called a compression wave, otherwise it is referred to as an expansion wave. The local velocity of sound is given by

$$a = \sqrt{\frac{\gamma p}{\rho}}$$

Since the changes across each infinitesimal wave are isentropic, from (5.3)

$$\frac{p}{p_r} = \left(\frac{a}{a_r}\right)^{\frac{2\gamma}{\gamma-1}}$$

$$\text{or } a = a_r \left(\frac{p}{p_r}\right)^{\frac{\gamma-1}{2\gamma}} \tag{5.10}$$

where a_r is the speed of sound ahead of the wave, where fluid velocity is V_r. Now from (5.4) across each infinitesimal wave.

$$da = \frac{\gamma-1}{2} dV$$

Integration along the wave gives

$$a = \frac{\gamma-1}{2} V + C_1 \qquad (C_1 \text{ is constant})$$

When $a = a_r$, $V = V_r$

Therefore $C_1 = a_r - \frac{\gamma-1}{2} V_r$

and $\frac{a}{a_r} - 1 = \frac{\gamma-1}{2} \left(\frac{V}{a_r} - \frac{V_r}{a_r}\right)$

Substitution for $\frac{a}{a_r}$ from (5.10) gives the following expression for the fluid velocity at a point in the wave as a function of the local pressure p

$$\frac{V}{a_r} = \frac{2}{\gamma-1}\left[\left(\frac{p}{p_r}\right)^{\frac{\gamma-1}{2\gamma}} - 1\right] + \frac{V_r}{a_r} \tag{5.11}$$

The local wave speed is given by

$$V_w = V + a$$

substituting for V and a from (5.10) and (5.11) gives

$$\frac{V_w}{a_r} = \frac{V_r}{a_r} + \left\{\frac{\gamma+1}{\gamma-1}\left(\frac{p}{p_r}\right)^{\frac{\gamma-1}{2\gamma}} - \frac{2}{\gamma-1}\right\} \tag{5.12}$$

Equations (5.10), (5.11) and (5.12) show that the speed of sound, the speed of the fluid and the wave speed increase with increase in pressure when the wave is moving in the same direction as the fluid particles. This means that the compression wave will steepen and the expansion wave will spread out as the wave propagates along the duct. (Fig. 5.4)

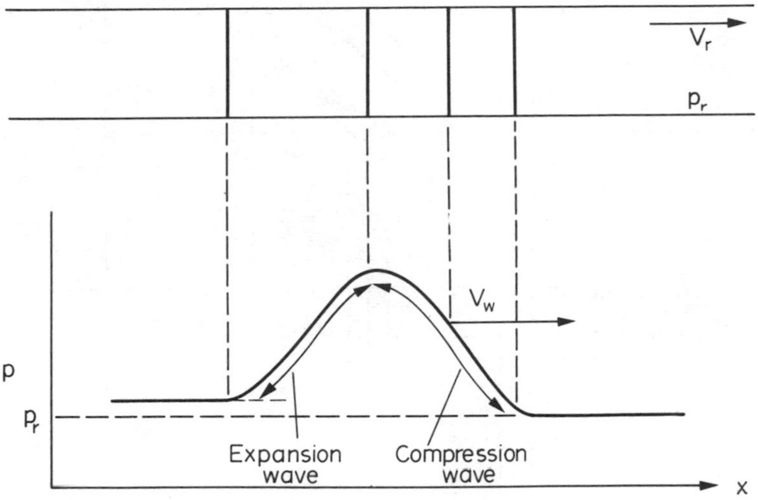

Fig. 5.3 Simple wave

At the time t_2 the compression wave becomes infinitely steep ($\frac{\partial p}{\partial x} = \infty$) at point A i.e., a small compression shock is formed. The present analysis then ceases to be valid because the effects of heat conduction and viscosity can no longer be neglected due to large gradients of the fluid properties. The shock grows in strength as the process continues. There is an entropy increase across the shock. The shock formation is a continuous process and the illustration given in Fig. 5.4 is a stepwise approximation to the actual process. Note that if the isentropic analysis were valid after t_2 then the wave would topple over to form the profile shown at t_3 by the dashed line. This is physically invalid since it implies that at the same time and location the fluid has three different values of any of the fluid properties. (The changes in wave form are also observed in gravity waves formed on liquid surfaces. However in this case the compression waves cannot give a sustained shock wave, but topple over thus forming a surf).

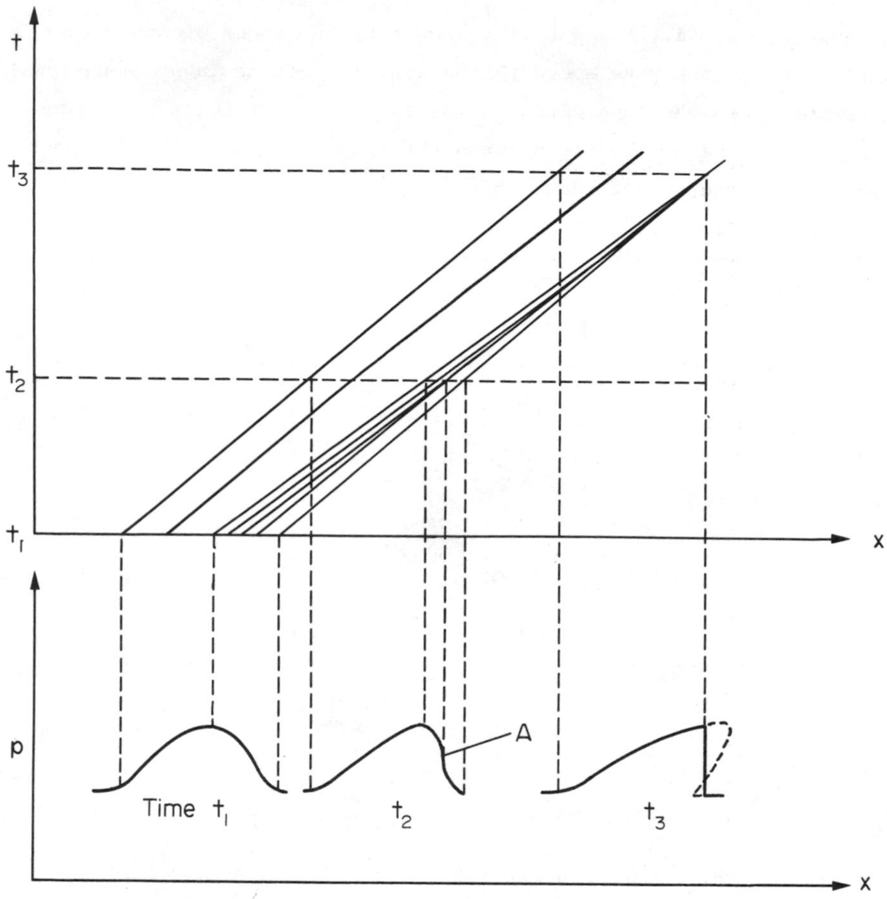

Fig. 5.4 Steepening of a compression wave and spreading of an expansion wave.

5.2.2 Waves of both families - Method of characteristics

In (5.1) it was shown that for infinitesimal disturbances

$$da = \pm \frac{\gamma-1}{2}\, dV \tag{5.13}$$

across a wave propagating with velocity

$$\left(\frac{dx}{dt}\right)_{wave} = V_w = V \pm a \tag{5.14}$$

The sign depends on the direction of propagation of the wave relative to the direction of the flow velocity, the plus sign is for the wave propagating in the same direction as the flow velocity and the minus sign for the wave propagating in the opposite direction to the flow velocity. If the fluid is flowing from left to right, and if the flow is subsonic, the + sign is for a wave propagating to the right (right-running wave) and the - sign for a wave propagating to the left (left-running wave). Integration of equation (5.13) gives, for right-running waves

$$(V_w = V + a), \quad \frac{a}{a_r} = \frac{\gamma-1}{2} \frac{V}{a_r} + C_1 \tag{5.15}$$

for left-running waves

$$(V_w = V - a), \quad \frac{a}{a_r} = -\left(\frac{\gamma-1}{2}\right) \frac{V}{a_r} + C_2 \tag{5.16}$$

where a_r is a reference speed, usually taken as the sonic speed in the fluid prior to the passage of the waves, and C_1 and C_2 are constants. Plotting $\frac{a}{a_r}$ versus $\frac{V}{a_r}$ for different values of C_1 and C_2, equations (5.15) and (5.16) give two families of straight lines (Fig. 5.5).

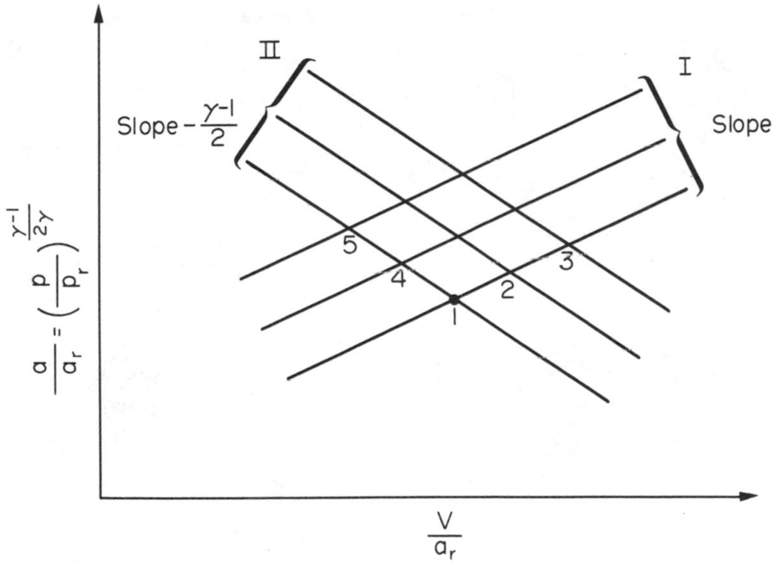

Fig. 5.5 State diagram

Fig. (5.5) is called the state diagram and the families of lines I and II are called the state characteristics. If the fluid is initially at state 1, then a right-running simple wave can change its state along line I passing through 1. Similarly a left-running simple wave can change its state along line II passing through 1. Thus with simple wave systems, the different states of the gas will be represented on a single line in the state diagram. (Note that for isentropic flow, with given initial conditions a_r, p_r, the state of the

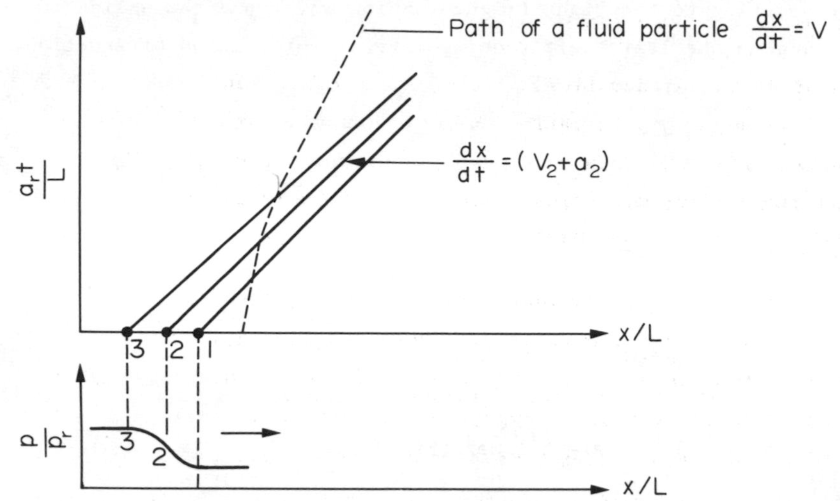

Fig. 5.6 Position Diagram

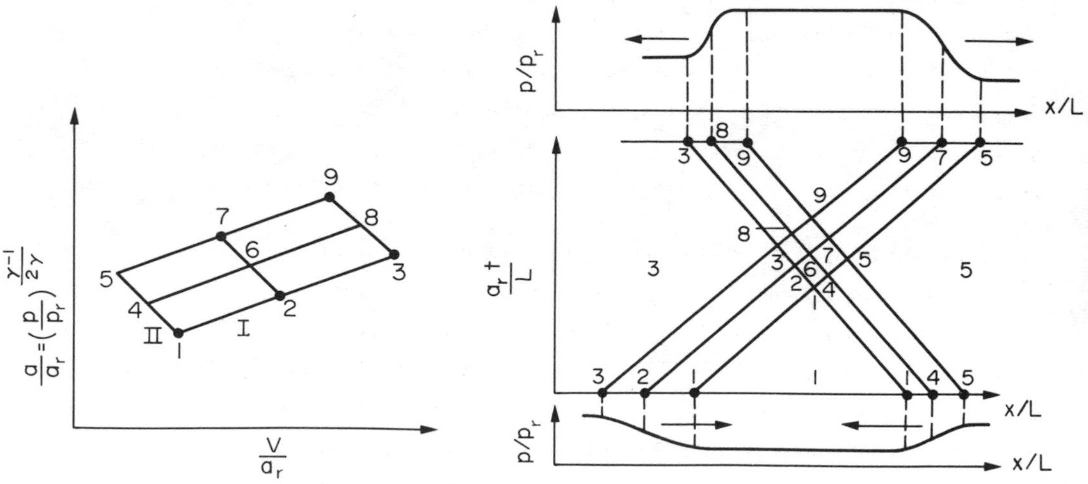

Fig. 5.7 Crossing of waves

fluid is defined by the sonic speed a.

The path of the waves and fluid particles can conveniently be represented on an (x-t) diagram. This is called the position diagram. The families of lines representing the path of right-running waves and left-running waves are called physical characteristics. Fig. (5.6) illustrates the position diagram, for a right-running simple wave.

Values of x and t are non-dimensionalized by using a reference length L (e.g. length of the duct) and the reference speed of sound a_r. Therefore starting with known initial conditions construction of the two diagrams gives a solution to the subsequent wave propagation, i.e. the wave positions and the fluid velocity and state at any given time can be found. Consider now the case where waves are propagating from both directions into region 1, Fig. (5.7).

Before the waves meet the state changes across the right-running wave are given by characteristics 1 - 3 in the state diagram. Similarly the state changes across the left-running wave are given by characteristic 1 - 5 in the state diagram. When the waves meet, there is a region of wave interaction and after the waves have crossed the state of the fluid is denoted by point 9. The state point 9 can be found by noting that point 9 must be on the II characteristic through 3 and the I characteristic through 5 and therefore lies on the point of intersection of the two characteristics. Within the interaction region, a state point in the state diagram can be found as follows:

Consider point 6, this point can be located by noting, that the state changes from 2 to 6 are caused by left-running waves only, since 2 - 6 is on a right-running characteristic in the position diagram. Similarly the states changes from 4 to 6 are caused by right-running characteristics only. Therefore point 6 must be on the point of intersection of a characteristic II through point 2 and a characteristic I through 4. Similarly points 7 and 8 can be located. Thus the state diagram can be constructed. The construction of the position diagram in the interaction region involves some degree of approximation. While for simple waves the slope of the characteristic in the position diagram is known, and remains constant as the wave propagates, in the interaction region the slope is continuously changing since values of a and V are being changed by wave interaction. Consider, e.g. characteristics 2 - 6 in the position diagram. Along 2 - 6 values of a and V are changing continuously by the passage of left-running waves, and hence the slope of the characteristic changes. If points 2 and 6 are sufficiently close then 2 - 6 can be drawn by assuming that it is straight and has a slope of

$$\frac{dx}{dt} = \left(\frac{V_2+V_6}{2} \right) + \left(\frac{a_2+a_6}{2} \right)$$

Similarly for 4 - 6. In this way the position diagram in the interaction region can be constructed step by step.

5.2.3 Boundaries

At the ends of the duct there are certain 'boundary' conditions that must be satisfied. For example at a closed end, the fluid velocity must remain zero. If waves propagate towards a closed end, the velocity induced by the waves must become zero at the closed end and this is achieved by reflected waves. Fig. (5.8) illustrates a compression wave moving towards the closed end. The state points on the wave are given by the characteristic 1-3 in the state diagram.

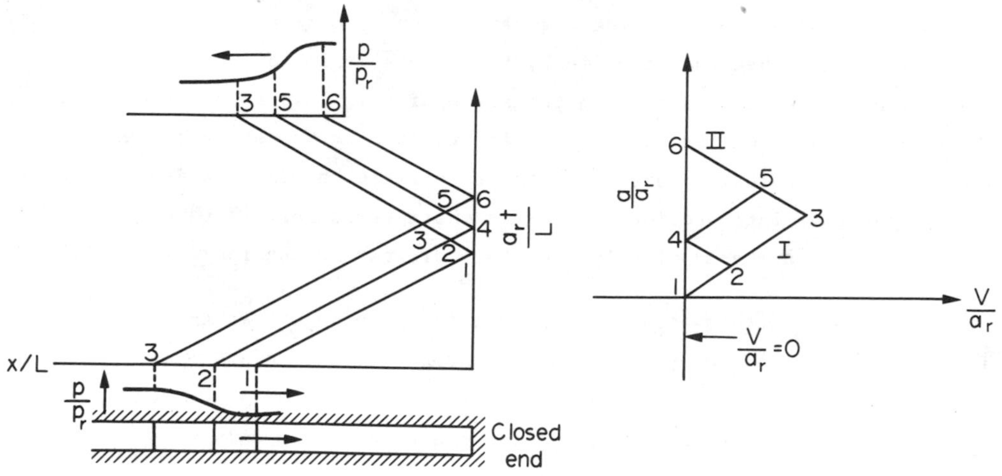

Fig. 5.8 Wave reflection from a closed end.

Velocity at the closed end (e.g. at points 4 and 6) must be zero and therefore left-running waves are reflected from the boundary to satisfy this condition. In the state diagram points 4 and 6 are the points of intersection of the II characteristics with $\frac{V}{a_r} = 0$ axis.

It can be seen that the compression wave reflects as another compression wave. In general waves are reflected from a closed end in like sense.

Near an open end there is complex unsteady three-dimensional motion as the wave propagates into the surroundings. However a reasonable approximation for subsonic flow is that pressure at the open end is equal to the pressure of the surroundings. From a theory of Helmholtz the inertia of the fluid outside the duct can be approximately taken into account by adding to the duct an equivalent length of $\frac{\pi d}{8}$, where d is the pipe diameter. Fig. (5.9) illustrates the reflection of a compression wave from the open end. The pressure of the surrounding is denoted by p_a and therefore points 4 and 6 in the state diagram, are the points of intersection of the II characteristics and the line $\left(\frac{p_a}{p_r}\right)^{\frac{\gamma-1}{2\gamma}}$ = constant.

It can be seen that the reflected wave is an expansion wave. In general waves reflect from a constant pressure boundary in an unlike sense.

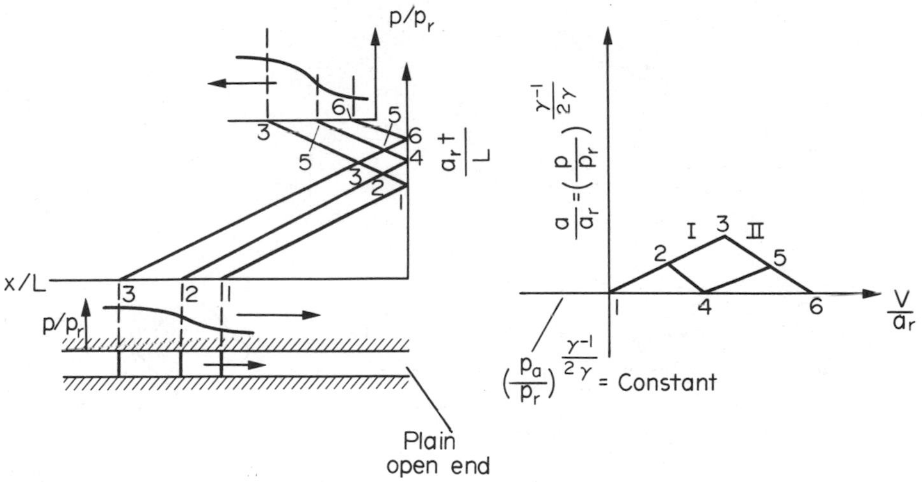

Fig. 5.9 Wave reflection at an open end.

Example: The system shown in Fig. (5.10) is initially at rest. The piston moves to the right with constant acceleration to a velocity V_3, after which the piston velocity is kept constant. The head of the reflected reaches the piston when the piston is moving with constant velocity.

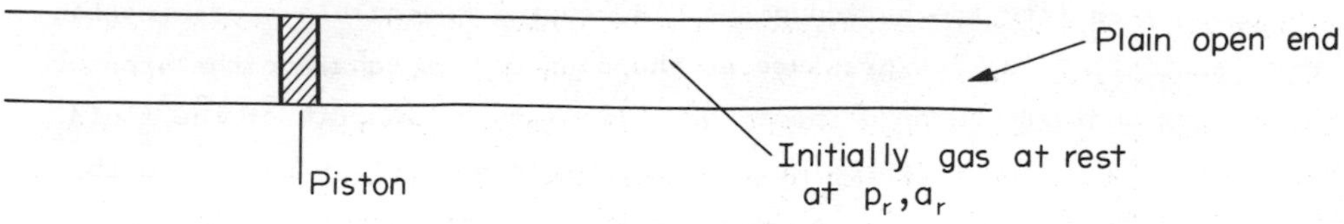

Fig. 5.10

The wave diagrams are shown in Fig. (5.11) and are self explanatory. The minimum pressure reached in the pipe (p_{min}) corresponds to the value at point 15, i.e.,

$$\frac{p_{min}}{p_r} = \left(\frac{a_{15}}{a_r} \right)^{\left(\frac{2\gamma}{\gamma-1} \right)}$$

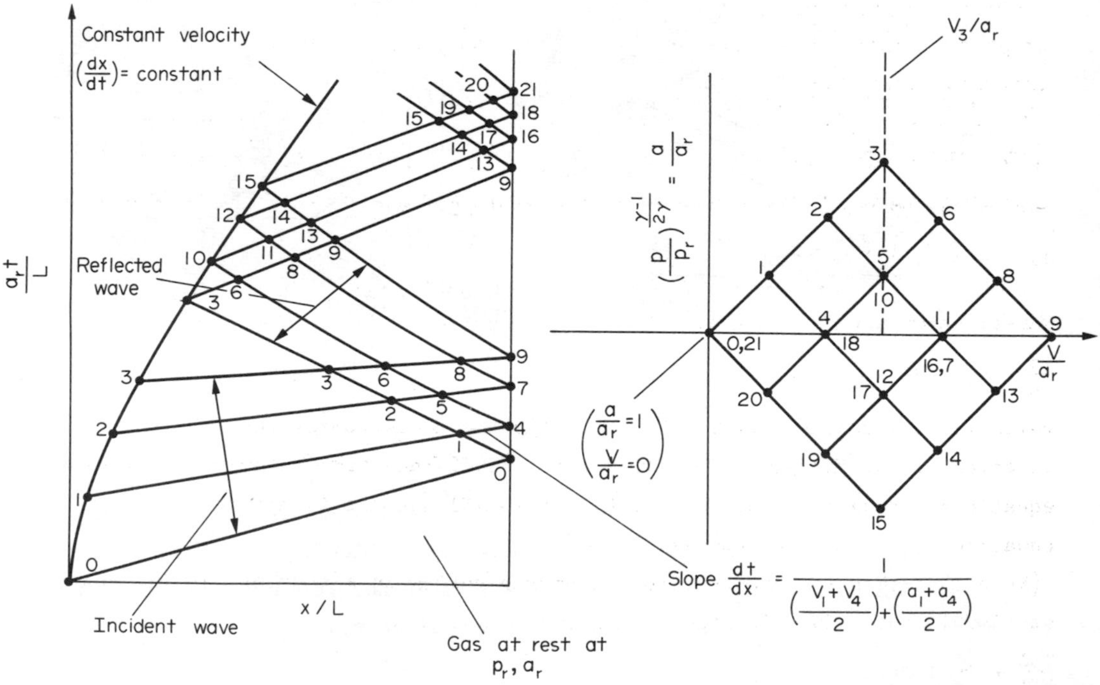

Fig. 5.11

5.3 Mathematical theory of one-dimensional homentropic unsteady flow

The results of the preceding sections can be obtained from the solution of the basic conservation equations derived in chapter 1. For simplicity, unsteady flow in a duct of constant area will be considered. Also it will be assumed that the entropy of all the fluid particles is the same and remains constant. That is, the fluid initially in the pipe and any fluid which enters the pipe have the same entropy. Therefore there must be no friction or heat transfer, otherwise the entropy of the fluid particles cannot remain constant. Such a flow is called homentropic. From the continuity equation (1.1),

$$\frac{\partial \rho}{\partial t} + \frac{\partial}{\partial x}(\rho V) = 0 \tag{5.17}$$

For a pure substance $p = f(\rho, s)$.

Therefore for homentropic flow $a^2 = \left(\frac{\partial p}{\partial \rho}\right)_s = \frac{dp}{d\rho}$ (5.18)

The momentum equation (1.4) states that

$$\frac{1}{\rho}\frac{\partial p}{\partial x} + \frac{\partial V}{\partial t} + V\frac{\partial V}{\partial x} = 0 \tag{5.19}$$

The energy equation (1.12) gives, for constant value of the area A,

$$\frac{\partial}{\partial t}\left(u + \frac{V^2}{2}\right) + V\frac{\partial}{\partial x}\left(u + \frac{V^2}{2}\right) + \frac{1}{\rho}\left(V\frac{\partial p}{\partial x} + p\frac{\partial V}{\partial x}\right) = 0$$

Using equation (5.19), the above equation reduces to

$$\frac{\partial u}{\partial t} + V\frac{\partial u}{\partial x} + \frac{p}{\rho}\frac{\partial V}{\partial x} = 0. \tag{5.20}$$

For homentropic flow ds = 0 and equation (1.14) gives

$$du = -\,p d\left(\frac{1}{\rho}\right) = \frac{p}{\rho^2}\,d\rho$$

Substitution into (5.20) gives

$$\frac{\partial \rho}{\partial t} + \frac{\partial}{\partial x}(\rho V) = 0$$

which is identical to (5.17). Therefore for the homentropic flow, the energy equation is not independent of (5.17) and (5.19) and can be derived from these equations. Equations (5.17), (5.18) and (5.19) are the independent set of equations for the flow process.

(Note that this is also true for homentropic flow in a duct of varying cross-sectional area. In this case from (1.1)

$$\frac{\partial \rho A}{\partial t} + \frac{\partial}{\partial x}(\rho A V) = 0$$

The momentum equation (5.19) is valid for the case of varying area.

The energy equation (1.12) gives

$$\frac{\partial}{\partial t}\left(u + \frac{V^2}{2}\right) + V\frac{\partial}{\partial x}\left(u + \frac{V^2}{2}\right) + \frac{V}{\rho}\frac{\partial p}{\partial x} + \frac{p}{\rho A}\frac{\partial}{\partial x}(AV) = 0$$

Using (5.19), the above equation reduces to

$$\frac{\partial u}{\partial t} + V\frac{\partial u}{\partial x} + \frac{p}{\rho A}\frac{\partial}{\partial x}(AV) = 0$$

Substituting for $du = \frac{p}{\rho 2}\,d\rho$, the above equation gives

$A\frac{\partial \rho}{\partial t} + \frac{\partial}{\partial x}(\rho AV) = 0$ which is the continuity equation for the case $A = A(x)$

for which the energy equation was derived.)

5.3.1 Small disturbances (acoustic waves)

It is difficult to obtain general solutions for equations (5.17) and (5.19) mainly because they are non-linear. However the equations can be made linear by considering the case of small perturbations i.e. the case where there are only small variations in the fluid properties compared with the values in the undisturbed fluid. This treatment is called the "small perturbation analysis".

Let
$$V = \overline{V} + V'$$
$$\rho = \overline{\rho} + \rho'$$
$$a = \overline{a} + a'$$
and
$$p = \overline{p} + p'$$

where $\overline{V}$, $\overline{\rho}$ etc. are the steady flow quantities and are independant of x and t for homentropic flow in a duct of constant cross-sectional area. The perturbations V', ρ' etc., are infinitesimal so that their squares and products can be neglected. Substitution in equations (5.17) and (5.19) gives

$$(\overline{\rho} + \rho')\frac{\partial V'}{\partial x} + (\overline{V} + V')\frac{\partial \rho'}{\partial x} + \frac{\partial \rho'}{\partial t} = 0$$

$$(\overline{V} + V')\frac{\partial V'}{\partial x} + \frac{\partial V'}{\partial t} + \frac{1}{\overline{\rho} + \rho'}\frac{\partial p'}{\partial x} = 0$$

Now, from (5.18) $dp = a^2\,d\rho$, therefore

$$\frac{1}{\overline{\rho} + \rho'}\frac{\partial p'}{\partial x} = \frac{a^2}{\overline{\rho} + \rho'}\frac{\partial \rho'}{\partial x} = \left(\frac{\overline{a}^2 + 2a'\overline{a}}{\overline{\rho}}\right)\left(1 - \frac{\rho'}{\overline{\rho}} + \ldots\right)\left(\frac{\partial \rho'}{\partial x}\right)$$

$$= \frac{\overline{a}^2}{\overline{\rho}}\left(\frac{\partial \rho'}{\partial x}\right)$$

Also, in the absence of discontinuities $V' \frac{\partial \rho'}{\partial x}$, and $V' \frac{\partial V'}{\partial x}$ can be neglected compared with the other terms of the equations. Therefore the above equations reduce to

$$\bar{\rho} \frac{\partial V'}{\partial x} + \frac{D\rho'}{Dt} = 0 \tag{5.21}$$

$$\frac{DV'}{Dt} + \frac{\bar{a}^2}{\bar{\rho}} \frac{\partial \rho'}{\partial x} = 0 \tag{5.22}$$

when $\frac{D}{Dt}$ denotes the substantive derivative $\frac{\partial}{\partial t} + \bar{V} \frac{\partial}{\partial x}$. Elimination of ρ', by taking $\frac{\partial}{\partial x}$ of the terms in (5.21) and $\frac{D}{Dt}$ of the terms in (5.22), gives the following equation for V'.

$$\bar{a}^2 \frac{\partial^2 V'}{\partial x^2} = \frac{D^2 V'}{Dt^2} = \left(\frac{\partial}{\partial t} + \bar{V} \frac{\partial}{\partial x} \right)\left(\frac{\partial V'}{\partial t} + \bar{V} \frac{\partial V'}{\partial x} \right) \tag{5.23}$$

For $\bar{V} = 0$, the above equation reduces to the wave equation.

$$\bar{a}^2 \frac{\partial^2 V'}{\partial x^2} = \frac{\partial^2 V'}{\partial t^2}$$

which can be shown (e.g. by direct substitution) to have a solution,

$$V' = f_1(x - \bar{a}t) + f_2(x + \bar{a}t)$$

where f_1 and f_2 are arbitrary functions of their respective arguments. By analogy, a solution of (5.23) can be found, by letting $V' = f(x - V_w t)$ where V_w is to be determined. Substitution in (5.23) gives $V_w = \bar{V} \pm \bar{a}$. Therefore the solution is

$$V' = f_1 \left(x - (\bar{V} + \bar{a})t \right) + f_2 \left(x - (\bar{V} - \bar{a})t \right) \tag{5.24}$$

For the simple wave solution $V' = f_1(x - (\bar{V} + \bar{a})t)$, the perturbations in pressure and density can be found as follows: From (5.22)

$$\left(\frac{\partial}{\partial t} + \bar{V} \frac{\partial}{\partial x} \right) f_1 \left(x - (\bar{V} + \bar{a})t \right) + \frac{\bar{a}^2}{\bar{\rho}} \frac{\partial \rho'}{\partial x} = 0$$

or

$$- (\bar{V} + \bar{a}) f_1' + \bar{V} f_1' + \frac{\bar{a}^2}{\bar{\rho}} \frac{\partial \rho'}{\partial x} = 0$$

where the prime denotes differentiation with respect to the argument of f_1. Therefore

$$\frac{\partial \rho'}{\partial x} = \frac{\bar{\rho}}{\bar{a}} f_1'$$

or

$$\rho' = \frac{\bar{\rho}}{\bar{a}} f_1 = \frac{\bar{\rho}}{\bar{a}} V' \tag{5.25}$$

which agrees with (5.1). Also by $dp' = \bar{a}^2 d\rho'$ so that

$$p' = \bar{a}\bar{\rho}V' \tag{5.26}$$

which agrees with (5.2). For the general solution (5.24),

$$\frac{\bar{a}^2}{\bar{\rho}} \frac{\partial \rho'}{\partial x} = \frac{1}{\bar{\rho}} \frac{\partial p'}{\partial x} = \bar{a}(f_1' - f_2')$$

so that

$$\rho' = \frac{\bar{\rho}}{\bar{a}} \left[f_1(x - (\bar{V} + \bar{a})t) - f_2(x - (\bar{V} - \bar{a})t) \right] \tag{5.27}$$

and

$$p' = \bar{\rho}\bar{a} \left[f_1(x - (\bar{V} + \bar{a})t) - f_2(x - (\bar{V} - \bar{a})t) \right] . \tag{5.28}$$

Note that the functions f_1 and f_2 are arbitrary and can be chosen at will to construct different flow patterns. Consider now each function separately. Let $f_1(x - (\bar{V} + \bar{a})t) = f_1(X)$ where $X = x - (\bar{a} + \bar{V})t$. $f_1(X)$ defines the form of the function which may be prescribed at a given time, say $t = 0$. The form of the function remains unchanged with time as the wave defined by $f_1(X)$ propagates along the x axis, see Fig. 5.12.

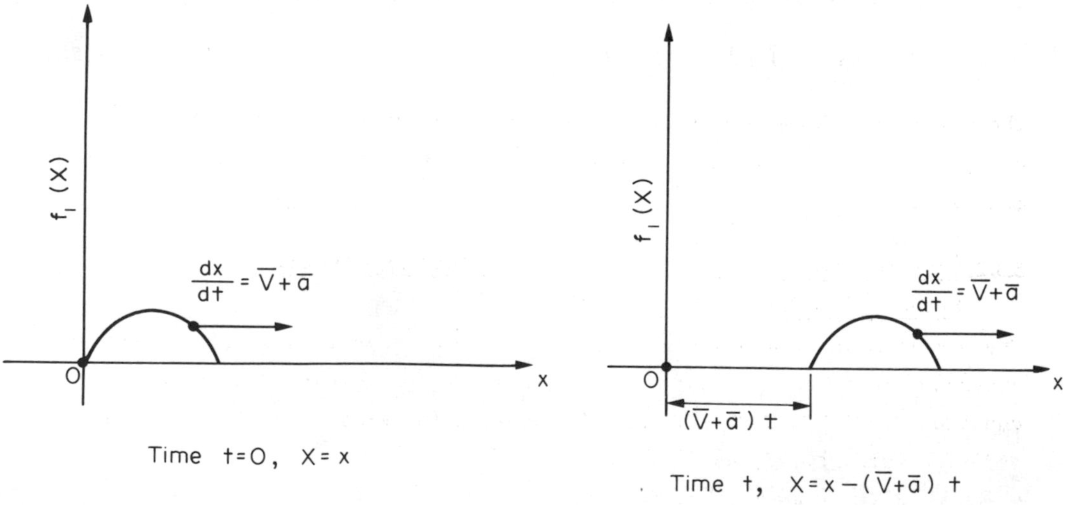

Fig. 5.12

The velocity of propagation $\frac{dx}{dt} = \bar{a} + \bar{V}$ is constant and is equal to the velocity of sound relative to the moving fluid. Similarly the function $f_2(x - (\bar{V} - \bar{a})t)$ describes a wave whose shape is fixed by the prescribed distribution of f_2 and this wave propagates with constant velocity $\frac{dx}{dt} = \bar{V} - \bar{a}$ along the axis, i.e. with velocity $-\bar{a}$ relative to the moving fluid. The complete solution of the equation is the sum of the two simple waves f_1 and f_2. Lines of constant f_1 and f_2 on the x,t diagram form two families of parallel straight lines of constant slope $\frac{dx}{dt} = \bar{V} \pm \bar{a}$, see Fig. 5.13. These lines are called characteristics of the wave equation.

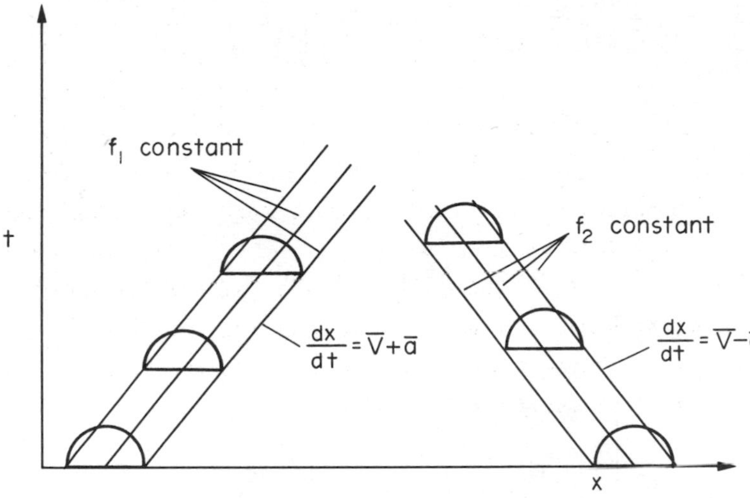

Fig. 5.13

The waves propagate along these lines with velocity $\bar{V} \pm \bar{a}$ and there is a change in the fluid properties across the lines (note analogy with the treatment in 5.1).

5.3.2. Unsteady flow with finite changes in fluid properties. Method of characteristics

The simple case of homentropic flow of a perfect gas in a duct of constant cross-sectional area, treated in 5.2, will be considered. The governing equations (5.17) and (5.19) can be expressed in terms of sonic speed, a, and flow velocity, V, as follows:

Since $a^2 = \frac{\gamma p}{\rho}$ then $2\,\frac{da}{a} = \frac{dp}{p} - \frac{d\rho}{\rho}$

Using (5.18), $2\ \frac{da}{a} = \frac{d\rho}{\rho}(\gamma - 1) = \frac{dp}{p}\left(\frac{\gamma - 1}{\gamma}\right)$

Substitution for derivatives of p and ρ in equations (5.17) and (5.19) gives

$$V\frac{\partial a}{\partial t} + \frac{\partial a}{\partial x} + \frac{\gamma - 1}{2}\ a\frac{\partial V}{\partial x} = 0 \qquad (5.29)$$

$$a\frac{\partial a}{\partial x} + \frac{\gamma - 1}{2}\left(V\frac{\partial V}{\partial x} + \frac{\partial V}{\partial t}\right) = 0 \qquad (5.30)$$

The basic idea of the method of characteristics is solution of a system of partial differential equations by transformation of this system into a system of ordinary differential equations along certain directions, called characteristic directions. Adding and subtracting (5.29) and (5.30) gives

$$\left[\frac{\partial a}{\partial t} + (V \pm a)\frac{\partial a}{\partial x}\right] \pm \frac{\gamma - 1}{2}\left[\frac{\partial V}{\partial t} + (V \pm a)\frac{\partial V}{\partial x}\right] = 0 \qquad (5.31)$$

Now for a function $\phi(x,t)$

$$d\phi = \frac{\partial \phi}{\partial x}dx + \frac{\partial \phi}{\partial t}dt.$$

Therefore the rate of change of ϕ with time is

$$\frac{d\phi}{dt} = \frac{\partial \phi}{\partial x}\left(\frac{dx}{dt}\right) + \frac{\partial \phi}{\partial t}$$

which depends on the direction $\left(\frac{dx}{dt}\right)$ along which $\frac{d\phi}{dt}$ is determined, see Fig. 5.14.

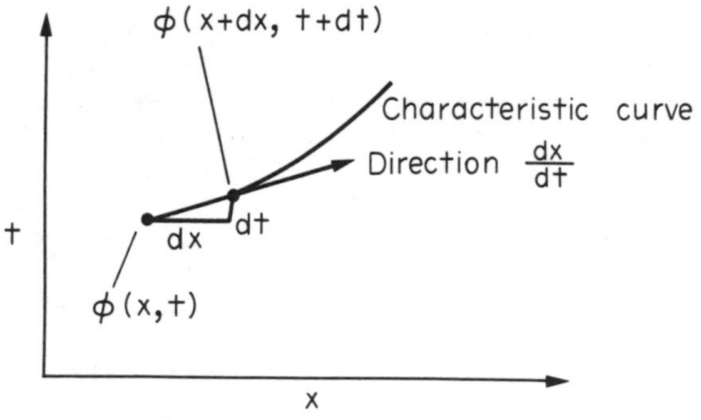

Fig. 5.14

In equation (5.31), each dependent variable V or a is differentiated along a direction in the x,t plane of the given slope. Therefore from equation (5.31)

$$\frac{da}{dt} \pm \frac{\gamma - 1}{2}\frac{dV}{dt} = 0 \text{ along the direction } \frac{dx}{dt} = V \pm a \qquad (5.32)$$

$$\text{i.e., } a \pm \frac{\gamma - 1}{2}V = \text{constant} \qquad (5.33)$$

along $\frac{dx}{dt} = V \pm a$ (5.34)

The families of lines given by (5.33) on the V, a plane are known as the state characteristics (Fig. 5.5) and the families of curves given by 5.34 on the x,t plane are known as the physical characteristics. Also equations 5.34 are known as the direction conditions (since these equations give the directions of the physical characteristics) and equations 5.33 are referred to as compatability conditions. (Note that equations 5.32 are consistent with equations (5.13) and (5.14) since moving along the physical characteristics of one family is equivalent to crossing the characteristics of the opposite family. Therefore

$da = \pm \frac{\gamma - 1}{2} dV$ across $(\frac{dx}{dt}) = V \pm a)$.

The utilization of the above results for solution of unsteady flow problems (by simultaneous construction of the state (V,a) and position (x,t) diagrams), and treatment of some simple boundaries was described in section 5.2.2. and 5.2.3

EXERCISES

1. A pipe contains air at 2 atmospheres and 290 K. Suddently the end blows off, dropping the pressure to 1 atmosphere. At what velocity does the air leave the open end?

Answer: 161 m/s

2. Water is flowing at 3 m/s along a pipe when a valve at exit is suddenly shut. Calculate the pressure rise at the valve, taking β (bulk modulus) as 22.3×10^5 kN/m^2 and ρ (density) as 1000 kg/m^3.

Why isn't the pressure rise the same after closing the valve slowly in practice?

Answer: 4485 kN/m^2.

3. The pressure p at the engine end of an exhaust pipe is given in kN/m^2 by

$p = 100 + 40 \sin (\frac{1}{8} \pi t)$

where t is the time in milliseconds measured from exhaust port opening. The exhaust pipe is 2m long and at time t = 0 the temperature of the gas in the pipe is 70°C, the pressure is 100 kN/m^2 and the particle velocity is zero. The other end of the pipe is open to the atmosphere at 100 kN/m^2.

Construct the velocity-time diagram at the open end for one cycle. Neglect the effect of reflected waves in constructing the position diagram. Assume $\gamma = 1.4$.

4. A tightly fitting piston can slide freely inside a pipe of uniform cross-sectional area. Initially the piston is at rest and it subsequently accelerates towards an open end of the pipe. The acceleration ceases when the velocity of the piston is V_o, after which its velocity remains constant. Initially the gas in the pipe is stationary and has uniform properties, the speed of sound being a_o. Treat the gas as perfect with ratio of specific heat capacities γ.

The first wave incident on the open end reflects to give a wave R, whose head meets the piston when the latter is no longer accelerating. Find, in terms of a_o, V_o and γ, the speed of the head of the wave R just before it meets the piston and the speed of the tail of R just after the reflection process at the open end. It may be assumed that no shock wave is formed.

Determine also the ratio of the initial pressure in the pipe to the pressure at the face of the piston just after reflection of wave R from the piston.

The rate of change of velocity magnitude with sound speed across the simple wave is $2/(\gamma - 1)$.

<u>Answer</u>: $a_o + \frac{\gamma - 3}{2}\,{}_o$, $a_o - 2V_o$, $\left(1 - \frac{\gamma - 1}{2}\frac{V_o}{a_o}\right)^{\frac{-2\gamma}{\gamma - 1}}$

5. A Pitot tube is idealized by taking it to be a duct of constant cross-sectional area with one end closed. It is placed in a pulsating flow, and the variation of the speed of sound at the open end can be represented by a periodic square wave with period 8 ms and amplitude 0.5% of the mean speed of sound. Calculate the percentage variation of pressure at the closed end, and sketch a graph comparing the pressure variations at both ends.

Flow in the Pitot tube may be taken to be homentropic, and changes in the slope of characteristics in the physical (x,t) plane may be neglected. The length of the duct is 0.35 m, the mean temperature of the gas is 305 K, $\gamma = 1.4$, and $R = 287 J/kgK$. It may be assumed that the gas is at rest in the duct and at the mean pressure whenever the speed of sound changes at the open end.

Answer: 7% from the mean.

6. A cylinder of length L contains a perfect gas in which the pressure is p_1 and the sound speed is a_1. The end $x = 0$ is closed by a diaphragm, the reverse side of which is exposed to a reservoir at pressure less than p_1. The end $x = L$ is closed by a piston. The gas and piston are initially at rest and the diaphragm is suddenly removed at time $t = 0$. The piston is to be moved subsequently in such a way that no waves are reflected from it. Show that, during the arrival of the wave, the piston must be moved in accordance with

$$\frac{x}{L} = -\frac{2}{\gamma - 1}\left(\frac{a_1 t}{L}\right) + \frac{\gamma + 1}{\gamma - 1}\left(\frac{a_1 t}{L}\right)^{\frac{2}{\gamma+1}}$$

and find how it must subsequently be moved.

It may be assumed without proof that the solution of the equation

$$\frac{dx}{dt} = A\frac{x}{t} - B,$$

where A and B are constants, is

$$x = \frac{B}{A - 1}t + Ct^A,$$

where C is a constant of integration.

7. Show that, for homentropic one-dimensional unsteady flow of a perfect gas in a duct of constant area, the equations of conservation of mass and momentum may be expressed respectively as

$$\frac{\partial a}{\partial t} + V\frac{\partial a}{\partial x} + \left(\frac{\gamma - 1}{2}\right) a \frac{\partial V}{\partial x} = 0$$

and

$$\frac{\partial V}{\partial t} + V\frac{\partial V}{\partial x} + \left(\frac{2}{\gamma - 1}\right) a \frac{\partial a}{\partial x} = 0,$$

where x is the distance along the duct, t the time, a the sonic velocity and V the velocity of fluid particles. Given that the spatial pressure gradient is zero, show that

$V = \frac{x}{t}$ and obtain $\left(\frac{a}{a_o}\right)$ in terms of $\left(\frac{t}{t_o}\right)$,

where a_o is the sonic velocity at time t_o. Hence obtain the equations of the characteristics (defined as $dx/dt = V \pm a$) in the form

$$\frac{x}{x_o} = f\left(\frac{t}{t_o}, \frac{a_o t}{x_o}\right),$$

where x_o is the value of x on each characteristic at time t_o. Comment on the initial and boundary conditions required to produce this flow.

Answer: $\frac{a}{a_o} = \left(\frac{t_o}{t}\right)^{\frac{\gamma-1}{2}}$, $x = \frac{tx_o}{t_o} \pm \frac{2a_o t}{\gamma - 1}\left[1 - \left(\frac{t_o}{t}\right)^{\frac{\gamma-1}{2}}\right]$.

8. A long tube is connected to a constant pressure reservoir. Referring to the sketch assume that initially the end of the tube is closed and that the air is at rest throughout the system. The end of the tube is suddenly opened. Draw the position and state diagrams for the subsequent wave action in the pipe. Find the number of wave reflections from the open end before pressure and velocity at all sections of the tube arrive within 5% of their respective steady state values.

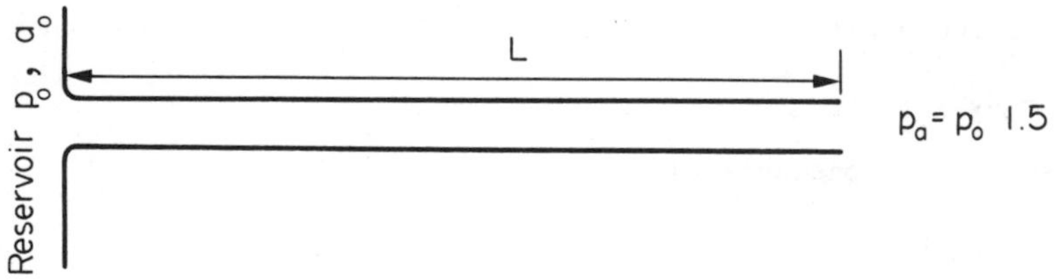

Answer: 3.

9. The system shown is initially at rest with the piston at x = 0. The piston moves to the left with constant acceleration ($\ddot{X} = d^2X/dZ^2 = c$ where $X = x/L$ and $Z = a_o t/L$) up to a speed $\bar{V}_1 = V_1/a_o$ (for which the flow remains subsonic) after which the velocity is constant. Show that the head of the reflected wave reaches the piston when the piston has travelled at distance

$$-\frac{\bar{V}_1^2}{2c} + \bar{V}_1 T + \bar{V}_1, \text{ where } T = \frac{1 + \bar{V}_1 - \bar{V}_1^2/2c}{1 + \bar{V}_1\left(\frac{1-\gamma}{2}\right)}$$

Assume (i) the speed of the reflected wave corresponds to that of the head of the reflected simple wave (ii) the head of the reflected wave reaches the piston when the piston is moving with constant velocity.

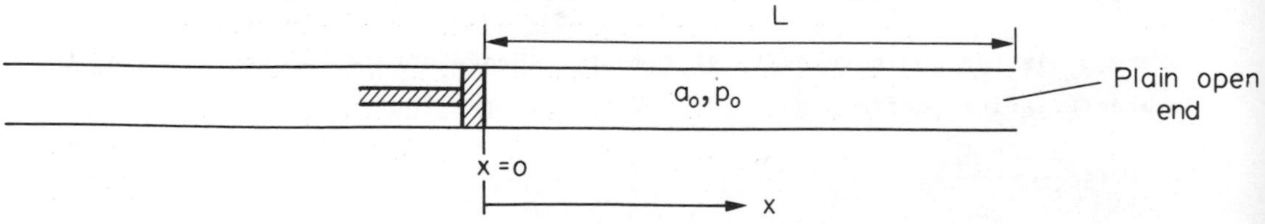

10. The cross-sectional area, A, of a duct varies with distance, x, along it according to the equation

$$A = A_o e^{kx}$$

where A_o and k are constants. Show that when small pressure disturbances, which vary sinusoidally with time at an angular frequency ω, are propagated through a fluid at rest in the duct, then the speed of propagation is

$$a\left\{1 - (ka/(2\omega))^2\right\}^{-\frac{1}{2}}$$

and the disturbances are attenuated by the factor

$$e^{-kx/2},$$

where a is the corresponding velocity of propagation of sound waves in a duct of uniform area, and $ka < 2\omega$. Neglect friction and heat transfer.

What happens if $ka > 2\omega$?

TABLE 1 ONE-DIMENSIONAL NORMAL SHOCK IN A PERFECT GAS

$\gamma = 1.400$

M_1	M_2	$\frac{p_2}{p_1}$	$\frac{T_2}{T_1}$	$\frac{p_{0_2}}{p_{0_1}}$	$\frac{p_{0_2}}{p_1}$
1.00	1.0000	1.000	1.000	1.000	1.893
1.02	0.9805	1.047	1.013	1.000	1.938
1.04	0.9620	1.095	1.026	1.000	1.984
1.06	0.9444	1.144	1.039	1.000	2.032
1.08	0.9277	1.194	1.052	0.999	2.082
1.10	0.9118	1.245	1.065	0.999	2.133
1.12	0.8966	1.297	1.078	0.998	2.185
1.14	0.8820	1.350	1.090	0.997	2.239
1.16	0.8682	1.403	1.103	0.996	2.294
1.18	0.8549	1.458	1.115	0.995	2.350
1.20	0.8422	1.513	1.128	0.993	2.408
1.22	0.8300	1.570	1.141	0.991	2.466
1.24	0.8183	1.627	1.153	0.988	2.526
1.26	0.8071	1.686	1.166	0.986	2.588
1.28	0.7963	1.745	1.178	0.983	2.650
1.30	0.7860	1.805	1.191	0.979	2.714
1.32	0.7760	1.866	1.204	0.976	2.778
1.34	0.7664	1.928	1.216	0.972	2.844
1.36	0.7572	1.991	1.229	0.968	2.912
1.38	0.7483	2.055	1.242	0.963	2.980
1.40	0.7397	2.120	1.255	0.958	3.049
1.42	0.7314	2.186	1.268	0.953	3.120
1.44	0.7235	2.253	1.281	0.948	3.191
1.46	0.7157	2.320	1.294	0.942	3.264
1.48	0.7083	2.389	1.307	0.936	3.338
1.50	0.7011	2.458	1.320	0.930	3.413
1.52	0.6941	2.529	1.334	0.923	3.489

NORMAL SHOCK

$\gamma = 1.400$

M_1	M_2	$\frac{p_2}{p_1}$	$\frac{T_2}{T_1}$	$\frac{p_{0_2}}{p_{0_1}}$	$\frac{p_{0_2}}{p_1}$
1.54	0.6874	2.600	1.347	0.917	3.567
1.56	0.6809	2.673	1.361	0.910	3.645
1.58	0.6746	2.746	1.374	0.903	3.724
1.60	0.6684	2.820	1.388	0.895	3.805
1.62	0.6625	2.895	1.402	0.888	3.887
1.64	0.6568	2.971	1.416	0.880	3.969
1.66	0.6512	3.048	1.430	0.872	4.053
1.68	0.6458	3.126	1.444	0.864	4.138
1.70	0.6405	3.205	1.458	0.856	4.224
1.72	0.6355	3.285	1.473	0.847	4.311
1.74	0.6305	3.366	1.487	0.839	4.399
1.76	0.6257	3.447	1.502	0.830	4.488
1.78	0.6210	3.530	1.517	0.822	4.578
1.80	0.6165	3.613	1.532	0.813	4.670
1.82	0.6121	3.698	1.547	0.804	4.762
1.84	0.6078	3.783	1.562	0.795	4.855
1.86	0.6036	3.870	1.577	0.786	4.950
1.88	0.5996	3.957	1.592	0.777	5.045
1.90	0.5956	4.045	1.608	0.767	5.142
1.92	0.5918	4.134	1.624	0.758	5.239
1.94	0.5880	4.224	1.639	0.749	5.338
1.96	0.5844	4.315	1.655	0.740	5.438
1.98	0.5808	4.407	1.671	0.730	5.539
2.00	0.5774	4.500	1.688	0.721	5.640
2.02	0.5740	4.594	1.704	0.712	5.743
2.04	0.5707	4.689	1.720	0.702	5.847
2.06	0.5675	4.784	1.737	0.693	5.952

NORMAL SHOCK

$\gamma = 1.400$

M_1	M_2	$\frac{p_2}{p_1}$	$\frac{T_2}{T_1}$	$\frac{p_{0_2}}{p_{0_1}}$	$\frac{p_{0_2}}{p_1}$
2.08	0.5643	4.881	1.754	0.684	6.058
2.10	0.5613	4.978	1.770	0.674	6.165
2.12	0.5583	5.077	1.787	0.665	6.274
2.14	0.5554	5.176	1.805	0.656	6.383
2.16	0.5525	5.277	1.822	0.646	6.493
2.18	0.5498	5.378	1.839	0.637	6.604
2.20	0.5471	5.480	1.857	0.628	6.716
2.22	0.5444	5.583	1.875	0.619	6.830
2.24	0.5418	5.687	1.892	0.610	6.944
2.26	0.5393	5.792	1.910	0.601	7.060
2.28	0.5368	5.898	1.929	0.592	7.176
2.30	0.5344	6.005	1.947	0.583	7.294
2.32	0.5321	6.113	1.965	0.575	7.412
2.34	0.5297	6.222	1.984	0.566	7.532
2.36	0.5275	6.331	2.002	0.557	7.652
2.38	0.5253	6.442	2.021	0.549	7.774
2.40	0.5231	6.553	2.040	0.540	7.897
2.42	0.5210	6.666	2.059	0.532	8.021
2.44	0.5189	6.779	2.079	0.523	8.145
2.46	0.5169	6.894	2.098	0.515	8.271
2.48	0.5149	7.009	2.118	0.507	8.398
2.5	0.5130	7.125	2.137	0.499	8.526
2.6	0.5039	7.720	2.238	0.460	9.181
2.7	0.4956	8.338	2.343	0.424	9.862
2.8	0.4882	8.980	2.451	0.389	10.569
2.9	0.4814	9.645	2.563	0.358	11.302

NORMAL SHOCK

$\gamma = 1.400$

M_1	M_2	$\frac{p_2}{p_1}$	$\frac{T_2}{T_1}$	$\frac{p_{0_2}}{p_{0_1}}$	$\frac{p_{0_2}}{p_1}$
3.0	0.4752	10.333	2.679	0.328	12.061
3.1	0.4695	11.045	2.799	0.301	12.846
3.2	0.4643	11.780	2.922	0.276	13.656
3.3	0.4596	12.538	3.049	0.253	14.492
3.4	0.4552	13.320	3.180	0.232	15.354
3.5	0.4512	14.125	3.315	0.213	16.242
3.6	0.4474	14.953	3.454	0.195	17.156
3.7	0.4439	15.805	3.596	0.179	18.095
3.8	0.4407	16.680	3.743	0.164	19.060
3.9	0.4377	17.578	3.893	0.151	20.051
4.0	0.4350	18.50	4.05	0.139	21.07
4.5	0.4236	23.46	4.88	0.092	26.54
5.0	0.4152	29.00	5.80	0.062	32.65
5.5	0.4090	35.12	6.82	0.042	39.41
6.0	0.4042	41.83	7.94	0.030	46.82
6.5	0.4004	49.12	9.16	0.021	54.86
7.0	0.3974	57.00	10.47	0.015	63.55
7.5	0.3949	65.46	11.88	0.011	72.89
8.0	0.3929	74.50	13.39	0.008	82.87
8.5	0.3912	84.13	14.99	0.006	93.49
9.0	0.3898	94.33	16.69	0.005	104.75
9.5	0.3886	105.13	18.49	0.004	116.66
∞	0.3780	∞	∞	0	∞

NORMAL SHOCK

$\gamma = 1.333$

M_1	M_2	$\frac{p_2}{p_1}$	$\frac{T_2}{T_1}$	$\frac{p_{0_2}}{p_{0_1}}$	$\frac{p_{0_2}}{p_1}$
1.00	1.0000	1.000	1.000	1.000	1.852
1.02	0.9805	1.046	1.011	1.000	1.896
1.04	0.9620	1.093	1.023	1.000	1.940
1.06	0.9443	1.141	1.034	1.000	1.986
1.08	0.9275	1.190	1.045	0.999	2.034
1.10	0.9114	1.240	1.055	0.999	2.083
1.12	0.8960	1.291	1.066	0.998	2.133
1.14	0.8814	1.342	1.077	0.997	2.184
1.16	0.8673	1.395	1.088	0.996	2.237
1.18	0.8539	1.448	1.098	0.994	2.291
1.20	0.8410	1.503	1.109	0.993	2.347
1.22	0.8286	1.558	1.120	0.991	2.403
1.24	0.8167	1.614	1.130	0.988	2.461
1.26	0.8053	1.671	1.141	0.985	2.520
1.28	0.7943	1.730	1.152	0.982	2.580
1.30	0.7837	1.788	1.163	0.979	2.641
1.32	0.7735	1.848	1.173	0.975	2.703
1.34	0.7637	1.909	1.184	0.971	2.767
1.36	0.7543	1.971	1.195	0.967	2.831
1.38	0.7451	2.033	1.206	0.962	2.897
1.40	0.7363	2.097	1.217	0.957	2.964
1.42	0.7279	2.161	1.227	0.952	3.032
1.44	0.7196	2.227	1.238	0.946	3.101
1.46	0.7117	2.293	1.250	0.940	3.171
1.48	0.7040	2.360	1.261	0.934	3.242

NORMAL SHOCK

$\gamma = 1.333$

M_1	M_2	$\frac{p_2}{p_1}$	$\frac{T_2}{T_1}$	$\frac{p_{o_2}}{p_{o_1}}$	$\frac{p_{o_2}}{p_1}$
1.50	0.6966	2.428	1.272	0.927	3.314
1.52	0.6894	2.497	1.283	0.921	3.388
1.54	0.6825	2.567	1.294	0.914	3.456
1.56	0.6757	2.638	1.306	0.906	3.537
1.58	0.6692	2.710	1.317	0.899	3.614
1.60	0.6629	2.783	1.329	0.891	3.692
1.62	0.6567	2.856	1.341	0.883	3.770
1.64	0.6508	2.931	1.352	0.875	3.850
1.66	0.6450	3.006	1.364	0.867	3.930
1.68	0.6394	3.083	1.376	0.858	4.012
1.70	0.6339	3.160	1.388	0.850	4.095
1.72	0.6286	3.238	1.400	0.841	4.179
1.74	0.6235	3.317	1.413	0.832	4.264
1.76	0.6185	3.397	1.425	0.823	4.349
1.78	0.6136	3.478	1.437	0.814	4.436
1.80	0.6089	3.560	1.450	0.804	4.524
1.82	0.6043	3.642	1.463	0.795	4.613
1.84	0.5998	3.726	1.475	0.786	4.703
1.86	0.5954	3.811	1.488	0.776	4.794
1.88	0.5912	3.896	1.501	0.766	4.886
1.90	0.5871	3.983	1.514	0.757	4.979
1.92	0.5830	4.070	1.527	0.747	5.073
1.94	0.5791	4.158	1.541	0.737	5.168
1.96	0.5753	4.247	1.554	0.727	5.264
1.98	0.5715	4.337	1.567	0.717	5.362
2.00	0.5679	4.428	1.581	0.708	5.460

NORMAL SHOCK

$\gamma = 1.333$

M_1	M_2	$\frac{p_2}{p_1}$	$\frac{T_2}{T_1}$	$\frac{p_{0_2}}{p_{0_1}}$	$\frac{p_{0_2}}{p_1}$
2.02	0.5644	4.520	1.595	0.698	5.559
2.04	0.5609	4.613	1.609	0.688	5.659
2.06	0.5575	4.707	1.623	0.678	5.760
2.08	0.5542	4.801	1.637	0.668	5.862
2.10	0.5510	4.897	1.651	0.658	5.965
2.12	0.5479	4.993	1.665	0.649	6.070
2.14	0.5448	5.091	1.680	0.639	6.175
2.16	0.5418	5.189	1.694	0.629	6.281
2.18	0.5389	5.288	1.709	0.619	6.388
2.20	0.5360	5.388	1.723	0.610	6.496
2.22	0.5332	5.489	1.738	0.600	6.606
2.24	0.5305	5.591	1.753	0.591	6.716
2.26	0.5278	5.694	1.768	0.581	6.827
2.28	0.5252	5.798	1.784	0.572	6.939
2.30	0.5227	5.902	1.799	0.563	7.053
2.32	0.5202	6.008	1.814	0.553	7.167
2.34	0.5177	6.114	1.830	0.544	7.282
2.36	0.5153	6.222	1.846	0.535	7.398
2.38	0.5130	6.330	1.862	0.526	7.516
2.40	0.5107	6.439	1.878	0.517	7.634
2.42	0.5084	6.550	1.894	0.508	7.753
2.44	0.5062	6.661	1.910	0.500	7.873
2.46	0.5041	6.773	1.926	0.491	7.995
2.48	0.5020	6.886	1.943	0.483	8.117
2.5	0.4999	6.999	1.959	0.474	8.240
2.6	0.4902	7.582	2.044	0.434	8.872
2.7	0.4814	8.188	2.132	0.396	9.528
2.8	0.4735	8.816	2.222	0.361	10.209

Table 1

NORMAL SHOCK

$\gamma = 1.333$

M_1	M_2	$\frac{p_2}{p_1}$	$\frac{T_2}{T_1}$	$\frac{p_{0_2}}{p_{0_1}}$	$\frac{p_{0_2}}{p_1}$
2.9	0.4662	9.468	2.316	0.328	10.915
3.0	0.4596	10.142	2.414	0.298	11.647
3.1	0.4535	10.839	2.514	0.271	12.403
3.2	0.4479	11.559	2.618	0.246	13.184
3.3	0.4428	12.302	2.724	0.223	13.990
3.4	0.4380	13.067	2.834	0.202	14.820
3.5	0.4337	13.856	2.947	0.183	15.676
3.6	0.4296	14.667	3.064	0.166	16.556
3.7	0.4259	15.501	3.183	0.150	17.462
3.8	0.4224	16.358	3.306	0.136	18.392
3.9	0.4191	17.238	3.432	0.124	19.347
4.0	0.4161	18.14	3.56	0.112	20.33
4.5	0.4037	23.00	4.26	0.070	25.60
5.0	0.3946	28.43	5.03	0.044	31.49
5.5	0.3877	34.42	5.89	0.028	38.01
6.0	0.3824	41.00	6.83	0.019	45.14
6.5	0.3783	48.14	7.85	0.013	52.90
7.0	0.3749	55.85	8.95	0.009	61.27
7.5	0.3722	64.14	10.13	0.006	70.27
8.0	0.3700	72.99	11.40	0.004	79.88
8.5	0.3681	82.42	12.74	0.003	90.12
9.0	0.3666	92.42	14.17	0.002	100.98
9.5	0.3652	102.99	15.68	0.002	112.46
∞	0.3534	∞	∞	0	∞

NORMAL SHOCK

$\gamma = 1.667$

M_1	M_2	$\frac{p_2}{p_1}$	$\frac{T_2}{T_1}$	$\frac{p_{0_2}}{p_{0_1}}$	$\frac{p_{0_2}}{p_1}$
1.00	1.0000	1.000	1.000	1.000	2.053
1.02	0.9806	1.051	1.020	1.000	2.105
1.04	0.9623	1.102	1.040	1.000	2.159
1.06	0.9450	1.155	1.059	1.000	2.215
1.08	0.9286	1.208	1.079	0.999	2.272
1.10	0.9131	1.263	1.098	0.999	2.331
1.12	0.8983	1.318	1.118	0.998	2.391
1.14	0.8843	1.375	1.137	0.997	2.453
1.16	0.8710	1.432	1.156	0.996	2.516
1.18	0.8583	1.491	1.176	0.995	2.581
1.20	0.8462	1.550	1.195	0.993	2.647
1.22	0.8347	1.611	1.214	0.991	2.715
1.24	0.8237	1.672	1.234	0.989	2.784
1.26	0.8132	1.735	1.253	0.987	2.854
1.28	0.8031	1.798	1.273	0.984	2.926
1.30	0.7934	1.863	1.292	0.981	2.999
1.32	0.7842	1.928	1.312	0.978	3.073
1.34	0.7753	1.995	1.332	0.975	3.149
1.36	0.7668	2.062	1.352	0.971	3.226
1.38	0.7586	2.131	1.372	0.967	3.304
1.40	0.7508	2.200	1.392	0.963	3.384
1.42	0.7432	2.271	1.412	0.958	3.465
1.44	0.7359	2.342	1.433	0.953	3.547
1.46	0.7289	2.415	1.453	0.949	3.630
1.48	0.7222	2.488	1.575	0.943	3.715
1.50	0.7157	2.563	1.495	0.938	3.801

NORMAL SHOCK

$\gamma = 1.667$

M_1	M_2	$\frac{p_2}{p_1}$	$\frac{T_2}{T_1}$	$\frac{p_{0_2}}{p_{0_1}}$	$\frac{p_{0_2}}{p_1}$
1.52	0.7094	2.638	1.516	0.932	3.888
1.54	0.7034	2.715	1.537	0.927	3.976
1.56	0.6975	2.792	1.559	0.921	4.066
1.58	0.6919	2.871	1.580	0.915	4.157
1.60	0.6864	2.950	1.602	0.909	4.249
1.62	0.6812	3.031	1.624	0.902	4.342
1.64	0.6761	3.112	1.646	0.896	4.437
1.66	0.6712	3.195	1.668	0.889	4.533
1.68	0.6664	3.278	1.691	0.882	4.630
1.70	0.6618	3.363	1.714	0.875	4.728
1.72	0.6573	3.448	1.736	0.868	4.827
1.74	0.6530	3.535	1.760	0.861	4.928
1.76	0.6488	3.622	1.783	0.854	5.030
1.78	0.6447	3.711	1.806	0.847	5.133
1.80	0.6407	3.800	1.830	0.839	5.237
1.82	0.6369	3.891	1.854	0.832	5.343
1.84	0.6332	3.982	1.878	0.824	5.449
1.86	0.6296	4.075	1.902	0.817	5.557
1.88	0.6261	4.168	1.927	0.809	5.666
1.90	0.6227	4.263	1.952	0.802	5.777
1.92	0.6194	4.358	1.977	0.794	5.888
1.94	0.6161	4.455	2.002	0.786	6.001
1.96	0.6130	4.552	2.027	0.778	6.115
1.98	0.6100	4.651	2.053	0.771	6.230
2.00	0.6070	4.750	2.079	0.763	6.346
2.02	0.6041	4.851	2.105	0.755	6.464

NORMAL SHOCK

$\gamma = 1.667$

M_1	M_2	$\frac{p_2}{p_1}$	$\frac{T_2}{T_1}$	$\frac{p_{0_2}}{p_{0_1}}$	$\frac{p_{0_2}}{p_1}$
2.04	0.6013	4.952	2.131	0.748	6.582
2.06	0.5986	5.055	2.157	0.740	6.702
2.08	0.5959	5.158	2.184	0.732	6.823
2.10	0.5933	5.263	2.211	0.724	6.946
2.12	0.5908	5.368	2.238	0.717	7.069
2.14	0.5884	5.475	2.266	0.709	7.194
2.16	0.5860	5.582	2.293	0.701	7.320
2.18	0.5836	5.691	2.321	0.694	7.447
2.20	0.5814	5.800	2.349	0.686	7.575
2.22	0.5791	5.911	2.378	0.679	7.704
2.24	0.5770	6.022	2.406	0.671	7.835
2.26	0.5748	6.135	2.435	0.663	7.967
2.28	0.5728	6.248	2.464	0.656	8.100
2.30	0.5708	6.363	2.493	0.649	8.234
2.32	0.5688	6.478	2.523	0.641	8.369
2.34	0.5669	6.595	2.553	0.634	8.506
2.36	0.5650	6.712	2.583	0.627	8.644
2.38	0.5632	6.831	2.613	0.620	8.782
2.40	0.5614	6.950	2.643	0.612	8.923
2.42	0.5596	7.071	2.674	0.605	9.064
2.44	0.5579	7.192	2.705	0.598	9.206
2.46	0.5563	7.315	2.736	0.591	9.350
2.48	0.5546	7.438	2.767	0.584	9.495
2.5	0.5530	7.563	2.799	0.578	9.641
2.6	0.5455	8.201	2.961	0.544	10.389
2.7	0.5388	8.863	3.128	0.512	11.166

NORMAL SHOCK

$\gamma = 1.667$

M_1	M_2	$\frac{p_2}{p_1}$	$\frac{T_2}{T_1}$	$\frac{p_{o_2}}{p_{o_1}}$	$\frac{p_{o_2}}{p_1}$
2.8	0.5327	9.551	3.302	0.482	11.973
2.9	0.5273	10.263	3.482	0.454	12.809
3.0	0.5223	11.001	3.668	0.427	13.675
3.1	0.5177	11.763	3.860	0.402	14.570
3.2	0.5136	12.551	4.058	0.379	15.495
3.3	0.5098	13.363	4.262	0.357	16.449
3.4	0.5063	14.201	4.473	0.336	17.433
3.5	0.5032	15.064	4.689	0.317	18.446
3.6	0.5002	15.951	4.912	0.299	19.489
3.7	0.4975	16.864	5.141	0.282	20.561
3.8	0.4950	17.801	5.376	0.266	21.662
3.9	0.4926	18.764	5.618	0.251	22.793
4.0	0.4905	19.75	5.87	0.237	23.95
4.5	0.4816	25.06	7.20	0.181	30.20
5.0	0.4752	31.00	8.68	0.140	37.17
5.5	0.4705	37.57	10.33	0.110	44.88
6.0	0.4668	44.75	12.12	0.088	53.33
6.5	0.4640	52.57	14.08	0.071	62.51
7.0	0.4617	61.00	16.19	0.058	72.43
7.5	0.4599	70.07	18.46	0.048	83.08
8.0	0.4584	79.76	20.88	0.040	94.46
8.5	0.4571	90.07	23.46	0.034	106.58
9.0	0.4560	101.01	26.20	0.029	119.44
9.5	0.4552	112.57	29.09	0.025	133.03
∞	0.4473	∞	∞	0	∞

TABLE 2 ONE-DIMENSIONAL ISENTROPIC FLOW OF A PERFECT GAS

$\gamma = 1.400$

M	$\frac{T}{T_o}$	$\frac{p}{p_o}$	$\frac{V}{\sqrt{c_p T_o}}$	$\frac{\dot{m}\sqrt{c_p T_o}}{A\, p_o}$	$\frac{A}{A^*}$	$\frac{I}{I^*}$
0.00	1.000	1.000	0.000	0.000	∞	∞
0.02	1.000	1.000	0.013	0.044	28.942	22.834
0.04	1.000	0.999	0.025	0.088	14.481	11.435
0.06	0.999	0.997	0.038	0.133	9.666	7.643
0.08	0.999	0.996	0.051	0.176	7.262	5.753
0.10	0.998	0.993	0.063	0.220	5.822	4.624
0.12	0.997	0.990	0.076	0.263	4.864	3.875
0.14	0.996	0.986	0.088	0.306	4.182	3.343
0.16	0.995	0.982	0.101	0.349	3.673	2.947
0.18	0.994	0.978	0.113	0.391	3.278	2.642
0.20	0.992	0.972	0.126	0.432	2.964	2.400
0.22	0.990	0.967	0.138	0.473	2.708	2.205
0.24	0.989	0.961	0.151	0.513	2.496	2.043
0.26	0.987	0.954	0.163	0.553	2.317	1.909
0.28	0.985	0.947	0.176	0.592	2.116	1.795
0.30	0.982	0.939	0.188	0.629	2.035	1.698
0.32	0.980	0.932	0.200	0.667	1.922	1.614
0.34	0.977	0.923	0.213	0.703	1.823	1.542
0.36	0.975	0.914	0.225	0.738	1.736	1.479
0.38	0.972	0.905	0.237	0.772	1.659	1.424
0.40	0.969	0.896	0.249	0.806	1.590	1.375
0.42	0.966	0.886	0.261	0.838	1.529	1.332
0.44	0.963	0.876	0.273	0.869	1.474	1.294
0.46	0.959	0.865	0.285	0.899	1.425	1.260
0.48	0.956	0.854	0.297	0.928	1.380	1.230
0.50	0.952	0.843	0.309	0.956	1.340	1.203
0.52	0.949	0.832	0.320	0.983	1.303	1.179

ISENTROPIC FLOW

$\gamma = 1.400$

M	$\frac{T}{T_o}$	$\frac{p}{p_o}$	$\frac{V}{\sqrt{c_p T_o}}$	$\frac{\dot{m}\sqrt{c_p T_o}}{A\, p_o}$	$\frac{A}{A*}$	$\frac{I}{I*}$
0.54	0.945	0.820	0.332	1.008	1.270	1.157
0.56	0.941	0.808	0.344	1.033	1.240	1.138
0.58	0.937	0.796	0.355	1.056	1.213	1.121
0.60	0.933	0.784	0.367	1.078	1.188	1.105
0.62	0.929	0.772	0.378	1.099	1.166	1.091
0.64	0.924	0.759	0.389	1.119	1.145	1.079
0.66	0.920	0.747	0.400	1.137	1.127	1.068
0.68	0.915	0.734	0.411	1.154	1.110	1.058
0.70	0.911	0.721	0.422	1.171	1.094	1.049
0.72	0.906	0.708	0.433	1.185	1.081	1.041
0.74	0.901	0.695	0.444	1.199	1.068	1.034
0.76	0.896	0.682	0.455	1.212	1.057	1.028
0.78	0.892	0.669	0.466	1.223	1.047	1.023
0.80	0.887	0.656	0.476	1.234	1.038	1.019
0.82	0.881	0.643	0.487	1.243	1.030	1.015
0.84	0.876	0.630	0.497	1.251	1.024	1.011
0.86	0.871	0.617	0.508	1.259	1.018	1.008
0.88	0.866	0.604	0.518	1.265	1.013	1.006
0.90	0.861	0.591	0.528	1.270	1.009	1.004
0.92	0.855	0.578	0.538	1.274	1.006	1.002
0.94	0.850	0.566	0.548	1.277	1.003	1.001
0.96	0.844	0.553	0.558	1.279	1.001	1.001
0.98	0.839	0.541	0.568	1.281	1.000	1.000
1.00	0.833	0.528	0.577	1.281	1.000	1.000
1.02	0.828	0.516	0.587	1.281	1.000	1.000
1.04	0.822	0.504	0.596	1.279	1.001	1.001
1.06	0.817	0.492	0.606	1.277	1.003	1.001

ISENTROPIC FLOW

$\gamma = 1.400$

M	$\frac{T}{T_o}$	$\frac{p}{p_o}$	$\frac{V}{\sqrt{c_pT_o}}$	$\frac{\dot{m}\sqrt{c_pT_o}}{A\,p_o}$	$\frac{A}{A^*}$	$\frac{I}{I^*}$
1.08	0.811	0.480	0.615	1.274	1.005	1.002
1.10	0.805	0.468	0.624	1.271	1.008	1.003
1.12	0.799	0.457	0.633	1.267	1.011	1.004
1.14	0.794	0.445	0.642	1.262	1.015	1.006
1.16	0.788	0.434	0.651	1.256	1.020	1.007
1.18	0.782	0.423	0.660	1.250	1.025	1.009
1.20	0.776	0.412	0.669	1.243	1.030	1.011
1.22	0.771	0.402	0.677	1.236	1.037	1.013
1.24	0.765	0.391	0.686	1.228	1.043	1.015
1.26	0.759	0.381	0.694	1.220	1.050	1.017
1.28	0.753	0.371	0.703	1.211	1.058	1.019
1.30	0.747	0.361	0.711	1.201	1.066	1.022
1.32	0.742	0.351	0.719	1.192	1.075	1.024
1.34	0.736	0.342	0.727	1.181	1.084	1.027
1.36	0.730	0.332	0.735	1.171	1.094	1.029
1.38	0.724	0.323	0.743	1.160	1.104	1.032
1.40	0.718	0.314	0.750	1.149	1.115	1.035
1.42	0.713	0.305	0.758	1.138	1.126	1.037
1.44	0.707	0.297	0.766	1.126	1.138	1.040
1.46	0.701	0.289	0.773	1.114	1.150	1.043
1.48	0.695	0.280	0.781	1.102	1.163	1.046
1.50	0.690	0.272	0.788	1.089	1.176	1.049
1.52	0.684	0.265	0.795	1.077	1.190	1.052
1.54	0.678	0.257	0.802	1.064	1.204	1.055
1.56	0.673	0.250	0.809	1.051	1.219	1.058
1.58	0.667	0.242	0.816	1.038	1.234	1.060

Table 2

ISENTROPIC FLOW

$\gamma = 1.400$

M	$\frac{T}{T_o}$	$\frac{p}{p_o}$	$\frac{V}{\sqrt{c_p T_o}}$	$\frac{\dot{m}\sqrt{c_p T_o}}{A\ p_o}$	$\frac{A}{A^*}$	$\frac{I}{I^*}$
1.60	0.661	0.235	0.823	1.025	1.250	1.063
1.62	0.656	0.228	0.830	1.011	1.267	1.066
1.64	0.650	0.222	0.836	0.998	1.284	1.069
1.66	0.645	0.215	0.843	0.985	1.301	1.072
1.68	0.639	0.209	0.849	0.971	1.319	1.075
1.70	0.634	0.203	0.856	0.958	1.338	1.079
1.72	0.628	0.197	0.862	0.944	1.357	1.082
1.74	0.623	0.191	0.869	0.931	1.376	1.085
1.76	0.617	0.185	0.875	0.917	1.397	1.088
1.78	0.612	0.179	0.881	0.904	1.418	1.091
1.80	0.607	0.174	0.887	0.890	1.439	1.094
1.82	0.602	0.169	0.893	0.877	1.461	1.096
1.84	0.596	0.164	0.899	0.863	1.484	1.099
1.86	0.591	0.159	0.904	0.850	1.507	1.102
1.88	0.586	0.154	0.910	0.837	1.531	1.105
1.90	0.581	0.149	0.916	0.824	1.555	1.108
1.92	0.576	0.145	0.921	0.811	1.580	1.111
1.94	0.571	0.140	0.927	0.798	1.606	1.114
1.96	0.566	0.136	0.932	0.785	1.633	1.117
1.98	0.561	0.132	0.938	0.772	1.660	1.120
2.00	0.556	0.1278	0.943	0.759	1.687	1.123
2.02	0.551	0.1239	0.948	0.747	1.716	1.126
2.04	0.546	0.1201	0.953	0.734	1.745	1.128
2.06	0.541	0.1164	0.958	0.722	1.775	1.131
2.08	0.536	0.1128	0.963	0.709	1.806	1.134

ISENTROPIC FLOW

$\gamma = 1.400$

M	$\frac{T}{T_o}$	$\frac{p}{p_o}$	$\frac{V}{\sqrt{c_p T_o}}$	$\frac{\dot{m}\sqrt{c_p T_o}}{A\, p_o}$	$\frac{A}{A^*}$	$\frac{I}{I^*}$
2.10	0.531	0.1094	0.968	0.697	1.837	1.137
2.12	0.527	0.1060	0.973	0.685	1.869	1.139
2.14	0.522	0.1027	0.978	0.674	1.902	1.142
2.16	0.517	0.0996	0.983	0.662	1.935	1.145
2.18	0.513	0.0965	0.987	0.650	1.970	1.147
2.20	0.508	0.0935	0.992	0.639	2.005	1.150
2.22	0.504	0.0906	0.996	0.628	2.041	1.153
2.24	0.599	0.0878	1.001	0.617	2.078	1.155
2.26	0.495	0.0851	1.005	0.606	2.115	1.158
2.28	0.490	0.0825	1.010	0.595	2.154	1.160
2.30	0.486	0.0800	1.014	0.584	2.193	1.163
2.32	0.482	0.0775	1.018	0.574	2.233	1.165
2.34	0.477	0.0751	1.022	0.563	2.274	1.168
2.36	0.473	0.0728	1.027	0.553	2.316	1.170
2.38	0.469	0.0706	1.031	0.543	2.359	1.173
2.40	0.465	0.0684	1.035	0.533	2.403	1.175
2.42	0.461	0.0663	1.039	0.523	2.448	1.177
2.44	0.456	0.0643	1.043	0.514	2.494	1.180
2.46	0.452	0.0623	1.046	0.504	2.540	1.182
2.48	0.448	0.0604	1.050	0.495	2.588	1.184
2.5	0.444	0.0585	1.054	0.486	2.637	1.187
2.6	0.425	0.0501	1.072	0.442	2.896	1.198
2.7	0.407	0.0430	1.089	0.402	3.183	1.208
2.8	0.389	0.0368	1.105	0.366	3.500	1.218
2.9	0.373	0.0317	1.120	0.333	3.850	1.228
3.0	0.357	0.0272	1.134	0.303	4.235	1.237
3.1	0.342	0.0234	1.147	0.275	4.657	1.245

ISENTROPIC FLOW

$\gamma = 1.400$

M	$\frac{T}{T_o}$	$\frac{p}{p_o}$	$\frac{V}{\sqrt{c_p T_o}}$	$\frac{\dot{m}\sqrt{c_p T_o}}{A\, p_o}$	$\frac{A}{A^*}$	$\frac{I}{I^*}$
3.2	0.328	0.0202	1.159	0.250	5.121	1.253
3 3	0.315	0.0175	1.171	0.228	5.629	1.260
3.4	0.302	0.0151	1.182	0.207	6.184	1.268
3.5	0.290	0.0131	1.192	0.189	6.790	1.274
3.6	0.278	0.0114	1.201	0.172	7.450	1.281
3.7	0.268	0.0099	1.210	0.157	8.169	1.287
3.8	0.257	0.0086	1.219	0.143	8.951	1.292
3.9	0.247	0.0075	1.227	0.131	9.799	1.298
4.0	0.238	0,659E-02	1.234	0.120E 00	10.719	1.303
4.5	0.198	0.346E-02	1.266	0.773E-01	16.562	1.325
5.0	0.167	0.189E-02	1.291	0.512E-01	25.000	1.342
5.5	0.142	0.107E-02	1.310	0.347E-01	36.869	1.355
6.0	0.122	0.633E-03	1.325	0.241E-01	53.180	1.365
6.5	0.106	0.385E-03	1.337	0.170E-01	75.134	1.374
7.0	0.093	0.242E-03	1.347	0.123E-01	104.143	1.381
7.5	0.082	0.155E-03	1.355	0.903E-02	141.842	1.387
8.0	0.072	0.102E-03	1.362	0.674E-02	190.110	1.391
8.5	0.065	0.690E-04	1.368	0.510E-02	251.087	1.396
9.0	0.058	0.474E-04	1.372	0.392E-02	327.190	1.399
9.5	0.052	0.331E-04	1.377	0.304E-02	421.132	1.402
∞	0	0	1.414	0	∞	1.429

ISENTROPIC FLOW

$\gamma = 1.333$

M	$\frac{T}{T_o}$	$\frac{p}{p_o}$	$\frac{V}{\sqrt{c_p T_o}}$	$\frac{\dot{m}\sqrt{c_p T_o}}{A\,p_o}$	$\frac{A}{A^*}$	$\frac{I}{I^*}$
0.00	1.000	1.000	0.000	0.000	∞	∞
0.02	1.000	1.000	0.012	0.046	29.159	23.159
0.04	1.000	0.999	0.023	0.092	14.590	11.597
0.06	0.999	0.998	0.035	0.138	9.738	7.750
0.08	0.999	0.996	0.046	0.184	7.315	5.833
0.10	0.998	0.993	0.058	0.230	5.865	4.687
0.12	0.998	0.990	0.069	0.275	4.900	3.927
0.14	0.997	0.987	0.081	0.320	4.212	3.388
0.16	0.996	0.983	0.092	0.364	3.699	2.986
0.18	0.995	0.979	0.104	0.408	3.301	2.676
0.20	0.993	0.974	0.115	0.451	2.984	2.430
0.22	0.992	0.968	0.126	0.494	2.726	2.231
0.24	0.991	0.963	0.138	0.536	2.512	2.067
0.26	0.989	0.956	0.149	0.578	2.332	1.930
0.28	0.987	0.949	0.161	0.618	2.179	1.814
0.30	0.985	0.942	0.172	0.658	2.047	1.715
0.32	0.983	0.935	0.183	0.697	1.933	1.630
0.34	0.981	0.927	0.194	0.735	1.833	1.557
0.36	0.979	0.918	0.206	0.772	1.745	1.492
0.38	0.977	0.909	0.217	0.808	1.667	1.436
0.40	0.974	0.900	0.228	0.843	1.598	1.386
0.42	0.971	0.891	0.239	0.877	1.536	1.342
0.44	0.969	0.881	0.250	0.909	1.481	1.303
0.46	0.966	0.871	0.261	0.941	1.431	1.268
0.48	0.963	0.860	0.272	0.972	1.386	1.237

ISENTROPIC FLOW

$\gamma = 1.333$

M	$\frac{T}{T_o}$	$\frac{p}{p_o}$	$\frac{V}{\sqrt{c_p T_o}}$	$\frac{\dot{m}\sqrt{c_p T_o}}{A\, p_o}$	$\frac{A}{A^*}$	$\frac{I}{I^*}$
0.50	0.960	0.849	0.283	1.001	1.345	1.210
0.52	0.957	0.838	0.294	1.029	1.308	1.185
0.54	0.954	0.827	0.304	1.057	1.275	1.163
0.56	0.950	0.816	0.315	1.082	1.244	1.143
0.58	0.947	0.804	0.326	1.107	1.217	1.125
0.60	0.943	0.792	0.336	1.130	1.192	1.109
0.62	0.940	0.780	0.347	1.152	1.169	1.095
0.64	0.936	0.768	0.357	1.173	1.148	1.082
0.66	0.932	0.756	0.368	1.193	1.129	1.071
0.68	0.929	0.743	0.378	1.211	1.112	1.060
0.70	0.925	0.731	0.388	1.229	1.096	1.051
0.72	0.921	0.718	0.399	1.244	1.082	1.043
0.74	0.916	0.705	0.409	1.259	1.070	1.036
0.76	0.912	0.692	0.419	1.273	1.058	1.030
0.78	0.908	0.680	0.429	1.285	1.048	1.024
0.80	0.904	0.667	0.439	1.296	1.039	1.019
0.82	0.899	0.654	0.449	1.306	1.031	1.015
0.84	0.895	0.641	0.459	1.315	1.024	1.012
0.86	0.890	0.628	0.468	1.323	1.018	1.009
0.88	0.886	0.615	0.478	1.329	1.013	1.006
0.90	0.881	0.603	0.488	1.335	1.009	1.004
0.92	0.876	0.590	0.497	1.339	1.006	1.003
0.94	0.872	0.577	0.506	1.343	1.003	1.001
0.96	0.867	0.565	0.516	1.345	1.001	1.001
0.98	0.862	0.552	0.525	1.346	1.000	1.000

ISENTROPIC FLOW

$\gamma = 1.333$

M	$\frac{T}{T_o}$	$\frac{p}{p_o}$	$\frac{V}{\sqrt{c_p T_o}}$	$\frac{\dot{m}\sqrt{c_p T_o}}{A\ p_o}$	$\frac{A}{A^*}$	$\frac{I}{I^*}$
1.00	0.857	0.540	0.534	1.347	1.000	1.000
1.02	0.852	0.528	0.543	1.346	1.000	1.000
1.04	0.847	0.515	0.552	1.345	1.001	1.001
1.06	0.842	0.503	0.561	1.343	1.003	1.001
1.08	0.837	0.491	0.570	1.340	1.005	1.002
1.10	0.832	0.480	0.579	1.336	1.008	1.003
1.12	0.827	0.468	0.588	1.331	1.012	1.005
1.14	0.822	0.457	0.596	1.326	1.016	1.006
1.16	0.817	0.445	0.605	1.320	1.020	1.008
1.18	0.812	0.434	0.614	1.313	1.026	1.010
1.20	0.807	0.423	0.622	1.306	1.032	1.012
1.22	0.801	0.412	0.630	1.298	1.038	1.014
1.24	0.796	0.402	0.638	1.289	1.045	1.016
1.26	0.791	0.391	0.647	1.280	1.052	1.018
1.28	0.786	0.381	0.655	1.270	1.060	1.021
1.30	0.780	0.371	0.663	1.260	1.069	1.023
1.32	0.775	0.361	0.671	1.249	1.078	1.026
1.34	0.770	0.351	0.678	1.238	1.088	1.029
1.36	0.765	0.341	0.686	1.227	1.098	1.031
1.38	0.759	0.332	0.694	1.215	1.109	1.034
1.40	0.754	0.323	0.701	1.202	1.120	1.037
1.42	0.749	0.314	0.709	1.190	1.132	1.040
1.44	0.743	0.305	0.716	1.177	1.144	1.043
1.46	0.738	0.296	0.724	1.164	1.157	1.046
1.48	0.733	0.288	0.731	1.150	1.171	1.050

ISENTROPIC FLOW

$\gamma = 1.333$

M	$\frac{T}{T_o}$	$\frac{p}{p_o}$	$\frac{V}{\sqrt{c_p T_o}}$	$\frac{\dot{m}\sqrt{c_p T_o}}{A\, p_o}$	$\frac{A}{A^*}$	$\frac{I}{I^*}$
1.50	0.727	0.280	0.738	1.137	1.185	1.053
1.52	0.722	0.272	0.745	1.123	1.200	1.056
1.54	0.717	0.264	0.752	1.109	1.215	1.059
1.56	0.712	0.256	0.759	1.094	1.231	1.062
1.58	0.706	0.249	0.766	1.080	1.247	1.066
1.60	0.701	0.241	0.773	1.066	1.264	1.069
1.62	0.696	0.234	0.780	1.051	1.281	1.072
1.64	0.691	0.227	0.787	1.036	1.300	1.076
1.66	0.685	0.221	0.793	1.021	1.318	1.079
1.68	0.680	0.214	0.800	1.007	1.338	1.082
1.70	0.765	0.208	0.806	0.992	1.358	1.086
1.72	0.670	0.201	0.812	0.977	1.379	1.089
1.74	0.665	0.195	0.819	0.962	1.400	1.092
1.76	0.660	0.189	0.825	0.947	1.422	1.096
1.78	0.655	0.183	0.831	0.932	1.445	1.099
1.80	0.650	0.178	0.837	0.917	1.468	1.103
1.82	0.645	0.172	0.843	0.903	1.492	1.106
1.84	0.640	0.167	0.849	0.888	1.517	1.109
1.86	0.635	0.162	0.855	0.873	1.543	1.113
1.88	0.630	0.157	0.861	0.858	1.569	1.116
1.90	0.625	0.152	0.867	0.844	1.596	1.119
1.92	0.620	0.147	0.872	0.830	1.624	1.122
1.94	0.615	0.143	0.878	0.815	1.652	1.126
1.96	0.610	0.138	0.883	0.801	1.682	1.129
1.98	0.605	0.134	0.889	0.787	1.712	1.132

ISENTROPIC FLOW

$\gamma = 1.333$

M	$\frac{T}{T_o}$	$\frac{p}{p_o}$	$\frac{V}{\sqrt{c_p T_o}}$	$\frac{\dot{m}\sqrt{c_p T_o}}{A\,p_o}$	$\frac{A}{A^*}$	$\frac{I}{I^*}$
2.00	0.600	0.1296	0.894	0.773	1.743	1.136
2.02	0.595	0.1255	0.899	0.759	1.774	1.139
2.04	0.591	0.1216	0.905	0.745	1.807	1.142
2.06	0.586	0.1177	0.910	0.732	1.841	1.145
2.08	0.581	0.1140	0.915	0.718	1.875	1.148
2.10	0.577	0.1104	0.920	0.705	1.910	1.151
2.12	0.572	0.1069	0.925	0.692	1.947	1.155
2.14	0.567	0.1035	0.930	0.679	1.984	1.158
2.16	0.563	0.1002	0.935	0.666	2.022	1.161
2.18	0.558	0.0970	0.940	0.653	2.061	1.164
2.20	0.554	0.0939	0.945	0.641	2.101	1.167
2.22	0.549	0.0909	0.949	0.629	2.142	1.170
2.24	0.545	0.0880	0.954	0.617	2.184	1.173
2.26	0.540	0.0851	0.959	0.605	2.228	1.176
2.28	0.536	0.0824	0.963	0.593	2.272	1.179
2.30	0.532	0.0798	0.968	0.581	2.317	1.182
2.32	0.527	0.0772	0.972	0.570	2.364	1.185
2.34	0.523	0.0747	0.977	0.558	2.412	1.187
2.36	0.519	0.0723	0.981	0.547	2.460	1.190
2.38	0.515	0.0700	0.985	0.536	2.510	1.193
2.40	0.510	0.0678	0.989	0.526	2.562	1.196
2.42	0.506	0.0656	0.994	0.515	2.614	1.199
2.44	0.502	0.0635	0.998	0.505	2.668	1.201
2.46	0.498	0.0614	1.002	0.495	2.723	1.204
2.48	0.494	0.095	1.006	0.485	2.779	1.207
2.5	0.490	0.0575	1.010	0.475	2.837	1.210
2.6	0.470	0.0489	1.029	0.428	3.147	1.223

ISENTROPIC FLOW

$\gamma = 1.333$

M	$\frac{T}{T_o}$	$\frac{p}{p_o}$	$\frac{V}{\sqrt{c_p T_o}}$	$\frac{\dot{m}\sqrt{c_p T_o}}{A\, p_o}$	$\frac{A}{A^*}$	$\frac{I}{I^*}$
2.7	0.452	0.0415	1.047	0.385	3.494	1.235
2.8	0.434	0.0353	1.064	0.347	3.883	1.247
2.9	0.417	0.0300	1.080	0.312	4.319	1.258
3.0	0.400	0.0256	1.095	0.280	4.805	1.269
3.1	0.385	0.0218	1.109	0.252	5.346	1.279
3.2	0.370	0.0186	1.123	0.226	5.949	1.289
3.3	0.355	0.0159	1.135	0.204	6.618	1.298
3.4	0.342	0.0136	1.147	0.183	7.361	1.306
3.5	0.329	0.0117	1.158	0.165	8.184	1.315
3.6	0.317	0.0100	1.169	0.148	9.094	1.323
3.7	0.305	0.0086	1.179	0.133	10.100	1.330
3.8	0.294	0.0074	1.188	0.120	11.210	1.337
3.9	0.283	0.0064	1.197	0.108	12.432	1.344
4.0	0.273	0.553E-02	1.206	0.978E-01	13.778	1.350
4.5	0.229	0.273E-02	1.242	0.592E-01	22.733	1.377
5.0	0.194	0.140E-02	1.270	0.368E-01	36.634	1.399
5.5	0.166	0.749E-03	1.292	0.234E-01	57.605	1.416
6.0	0.143	0.415E-03	1.309	0.152E-01	88.435	1.429
6.5	0.124	0.238E-03	1.323	0.101E-01	132.704	1.440
7.0	0.109	0.141E-03	1.335	0.691E-02	194.929	1.449
7.5	0.096	0.860E-04	1.344	0.480E-02	280.720	1.457
8.0	0.086	0.538E-04	1.352	0.339E-02	396.951	1.463
8.5	0.077	0.344E-04	1.359	0.244E-02	551.940	1.468
9.0	0.069	0.225E-04	1.365	0.178E-02	755.644	1.473
9.5	0.062	0.150E-04	1.369	0.132E-02	1019.863	1.477
∞	0	0	1.414	0	∞	1.512

ISENTROPIC FLOW

$\gamma = 1.667$

M	$\frac{T}{T_o}$	$\frac{p}{p_o}$	$\frac{V}{\sqrt{c_p T_o}}$	$\frac{\dot{m}\sqrt{c_p T_o}}{A\, p_o}$	$\frac{A}{A^*}$	$\frac{I}{I^*}$
0.00	1.000	1.000	0.000	0.000	∞	∞
0.02	1.000	1.000	0.016	0.041	28.131	21.662
0.04	0.999	0.999	0.033	0.082	14.077	10.851
0.06	0.999	0.997	0.049	0.122	9.397	7.255
0.08	0.998	0.995	0.065	0.163	7.061	5.464
0.10	0.997	0.992	0.082	0.203	5.662	4.395
0.12	0.995	0.988	0.098	0.243	4.732	3.686
0.14	0.994	0.984	0.114	0.282	4.070	3.183
0.16	0.992	0.979	0.130	0.321	3.576	2.810
0.18	0.989	0.973	0.146	0.360	3.193	2.522
0.20	0.987	0.967	0.162	0.398	2.888	2.294
0.22	0.984	0.961	0.178	0.435	2.640	2.110
0.24	0.981	0.954	0.194	0.472	2.435	1.959
0.26	0.978	0.946	0.210	0.508	2.262	1.832
0.28	0.975	0.938	0.226	0.543	2.115	1.726
0.30	0.971	0.929	0.241	0.577	1.989	1.635
0.32	0.967	0.920	0.257	0.611	1.880	1.558
0.34	0.963	0.910	0.272	0.643	1.784	1.490
0.36	0.959	0.900	0.288	0.675	1.700	1.432
0.38	0.954	0.889	0.303	0.706	1.626	1.381
0.40	0.949	0.878	0.318	0.736	1.560	1.336
0.42	0.944	0.867	0.333	0.765	1.501	1.296
0.44	0.939	0.855	0.348	0.793	1.449	1.262
0.46	0.934	0.843	0.363	0.819	1.401	1.231
0.48	0.929	0.831	0.378	0.845	1.359	1.203

ISENTROPIC FLOW

$\gamma = 1.667$

M	$\frac{T}{T_o}$	$\frac{p}{p_o}$	$\frac{V}{\sqrt{c_p T_o}}$	$\frac{\dot{m}\sqrt{c_p T_o}}{A\, p_o}$	$\frac{A}{A^*}$	$\frac{I}{I^*}$
0.50	0.923	0.819	0.392	0.870	1.320	1.179
0.52	0.917	0.806	0.407	0.893	1.285	1.157
0.54	0.911	0.793	0.421	0.916	1.254	1.138
0.56	0.905	0.780	0.435	0.937	1.225	1.120
0.58	0.899	0.767	0.449	0.957	1.200	1.105
0.60	0.893	0.753	0.463	0.976	1.176	1.091
0.62	0.886	0.740	0.477	0.994	1.155	1.079
0.64	0.880	0.726	0.490	1.011	1.135	1.068
0.66	0.873	0.712	0.504	1.027	1.118	1.058
0.68	0.866	0.699	0.517	1.042	1.102	1.050
0.70	0.860	0.685	0.530	1.056	1.087	1.042
0.72	0.853	0.671	0.543	1.068	1.075	1.035
0.74	0.846	0.658	0.556	1.080	1.063	1.029
0.76	0.838	0.644	0.568	1.091	1.053	1.024
0.78	0.831	0.630	0.581	1.100	1.043	1.019
0.80	0.824	0.617	0.593	1.109	1.035	1.016
0.82	0.817	0.603	0.605	1.117	1.028	1.012
0.84	0.810	0.590	0.617	1.124	1.022	1.009
0.86	0.802	0.576	0.629	1.130	1.016	1.007
0.88	0.795	0.563	0.641	1.135	1.012	1.005
0.90	0.787	0.550	0.652	1.139	1.008	1.003
0.92	0.780	0.537	0.664	1.142	1.005	1.002
0.94	0.772	0.524	0.675	1.145	1.003	1.001
0.96	0.765	0.512	0.686	1.147	1.001	1.000
0.98	0.757	0.499	0.697	1.148	1.000	1.000

ISENTROPIC FLOW

$\gamma = 1.667$

M	$\frac{T}{T_o}$	$\frac{p}{p_o}$	$\frac{V}{\sqrt{c_p T_o}}$	$\frac{\dot{m}\sqrt{c_p T_o}}{A\,p_o}$	$\frac{A}{A^*}$	$\frac{I}{I^*}$
1.00	0.750	0.487	0.707	1.148	1.000	1.000
1.02	0.742	0.475	0.718	1.148	1.000	1.000
1.04	0.735	0.463	0.728	1.147	1.001	1.000
1.06	0.727	0.451	0.738	1.145	1.003	1.001
1.08	0.720	0.440	0.748	1.143	1.005	1.002
1.10	0.712	0.429	0.758	1.140	1.007	1.002
1.12	0.705	0.418	0.768	1.137	1.010	1.003
1.14	0.698	0.407	0.778	1.133	1.014	1.005
1.16	0.690	0.396	0.787	1.128	1.017	1.006
1.18	0.683	0.385	0.796	1.124	1.022	1.007
1.20	0.676	0.375	0.806	1.118	1.027	1.008
1.22	0.668	0.365	0.815	1.112	1.032	1.010
1.24	0.661	0.355	0.823	1.106	1.038	1.012
1.26	0.654	0.346	0.832	1.100	1.044	1.013
1.28	0.647	0.336	0.841	1.093	1.051	1.015
1.30	0.640	0.327	0.849	1.086	1.057	1.017
1.32	0.632	0.318	0.857	1.078	1.065	1.019
1.34	0.625	0.309	0.865	1.070	1.073	1.020
1.36	0.618	0.301	0.874	1.062	1.081	1.022
1.38	0.612	0.293	0.881	1.054	1.089	1.024
1.40	0.605	0.284	0.889	1.045	1.098	1.026
1.42	0.598	0.277	0.897	1.037	1.108	1.028
1.44	0.591	0.269	0.904	1.028	1.117	1.030
1.46	0.584	0.261	0.912	1.018	1.127	1.032
1.48	0.578	0.254	0.919	1.009	1.138	1.034
1.50	0.571	0.247	0.926	1.000	1.148	1.037
1.52	0.565	0.240	0.933	0.990	1.160	1.039

Table 2

ISENTROPIC FLOW

$\gamma = 1.667$

M	$\frac{T}{T_o}$	$\frac{p}{p_o}$	$\frac{V}{\sqrt{c_p T_o}}$	$\frac{\dot{m}\sqrt{c_p T_o}}{A\, p_o}$	$\frac{A}{A^*}$	$\frac{I}{I^*}$
1.54	0.558	0.233	0.940	0.980	1.171	1.041
1.56	0.552	0.226	0.947	0.971	1.183	1.043
1.58	0.546	0.220	0.953	0.961	1.195	1.045
1.60	0.539	0.214	0.960	0.951	1.208	1.047
1.62	0.533	0.208	0.966	0.941	1.220	1.049
1.64	0.527	0.202	0.972	0.931	1.234	1.051
1.66	0.521	0.196	0.979	0.921	1.247	1.053
1.68	0.515	0.191	0.985	0.910	1.261	1.055
1.70	0.509	0.185	0.991	0.900	1.275	1.057
1.72	0.503	0.180	0.997	0.890	1.290	1.059
1.74	0.498	0.175	1.002	0.880	1.305	1.061
1.76	0.492	0.170	1.008	0.870	1.320	1.063
1.78	0.486	0.165	1.014	0.859	1.336	1.065
1.80	0.481	0.160	1.019	0.849	1.352	1.067
1.82	0.475	0.156	1.025	0.839	1.368	1.069
1.84	0.470	0.151	1.030	0.829	1.385	1.071
1.86	0.464	0.147	1.035	0.819	1.402	1.073
1.88	0.459	0.143	1.040	0.809	1.419	1.075
1.90	0.454	0.139	1.045	0.799	1.437	1.077
1.92	0.449	0.135	1.050	0.789	1.455	1.079
1.94	0.443	0.131	1.055	0.779	1.474	1.081
1.96	0.438	0.127	1.060	0.769	1.492	1.083
1.98	0.433	0.124	1.065	0.760	1.512	1.085

ISENTROPIC FLOW

$\gamma = 1.667$

M	$\frac{T}{T_o}$	$\frac{p}{p_o}$	$\frac{V}{\sqrt{c_p T_o}}$	$\frac{\dot{m}\sqrt{c_p T_o}}{A\, p_o}$	$\frac{A}{A^*}$	$\frac{I}{I^*}$
2.00	0.428	0.1202	1.069	0.750	1.531	1.087
2.02	0.424	0.1168	1.074	0.740	1.551	1.088
2.04	0.419	0.1136	1.078	0.731	1.571	1.090
2.06	0.414	0.1104	1.083	0.721	1.592	1.092
2.08	0.409	0.1073	1.087	0.712	1.613	1.094
2.10	0.405	0.1043	1.091	0.703	1.634	1.095
2.12	0.400	0.1014	1.095	0.693	1.656	1.097
2.14	0.396	0.0986	1.099	0.684	1.678	1.099
2.16	0.391	0.0958	1.103	0.675	1.700	1.101
2.18	0.387	0.0932	1.107	0.666	1.723	1.102
2.20	0.383	0.0906	1.111	0.658	1.746	1.104
2.22	0.378	0.0881	1.115	0.649	1.769	1.105
2.24	0.374	0.0856	1.119	0.640	1.793	1.107
2.26	0.370	0.0833	1.123	0.632	1.818	1.109
2.28	0.366	0.0810	1.126	0.623	1.842	1.110
2.30	0.362	0.0788	1.130	0.615	1.867	1.112
2.32	0.358	0.0766	1.133	0.607	1.893	1.113
2.34	0.354	0.0745	1.137	0.598	1.918	1.115
2.36	0.350	0.0725	1.140	0.590	1.945	1.116
2.38	0.346	0.0705	1.144	0.582	1.971	1.118
2.40	0.342	0.0686	1.147	0.575	1.998	1.119
2.42	0.339	0.0668	1.150	0.567	2.025	1.121
2.44	0.335	0.0650	1.153	0.559	2.053	1.122
2.46	0.331	0.0632	1.156	0.552	2.081	1.123
2.48	0.328	0.0615	1.160	0.544	2.110	1.125

ISENTROPIC FLOW

$\gamma = 1.667$

M	$\frac{T}{T_o}$	$\frac{p}{p_o}$	$\frac{V}{\sqrt{c_p T_o}}$	$\frac{\dot{m}\sqrt{c_p T_o}}{A\ p_o}$	$\frac{A}{A*}$	$\frac{I}{I*}$
2.5	0.324	0.0599	1.163	0.537	2.139	1.126
2.6	0.307	0.0524	1.177	0.502	2.289	1.133
2.7	0.291	0.0459	1.190	0.469	2.450	1.139
2.8	0.277	0.0403	1.203	0.438	2.622	1.144
2.9	0.263	0.0355	1.214	0.409	2.805	1.150
3.0	0.250	0.0313	1.225	0.383	2.999	1.155
3.1	0.235	0.0276	0.358	0.358	3.205	1.159
3.2	0.226	0.0244	1.244	0.335	3.422	1.164
3.3	0.216	0.0217	1.252	0.314	3.653	1.168
3.4	0.206	0.0193	1.260	0.295	3.895	1.172
3.5	0.197	0.0172	1.268	0.277	4.151	1.175
3.6	0.188	0.0153	1.274	0.260	4.420	1.178
3.7	0.180	0.0137	1.281	0.244	4.703	1.182
3.8	0.172	0.0123	1.287	0.230	5.000	1.185
3.9	0.165	0.0110	1.293	0.216	5.311	1.187
4.0	0.158	0.991E-02	1.298	0.204E 00	5.637	1.190
4.5	0.129	0.598E-02	1.320	0.153E 00	7.503	1.201
5.0	0.107	0.376E-02	1.336	0.117E 00	9.792	1.209
5.5	0.090	0.245E-02	1.349	0.915E-01	12.552	1.216
6.0	0.077	0.164E-02	1.359	0.725E-01	15.827	1.221
6.5	0.066	0.113E-02	1.367	0.584E-01	19.666	1.225
7.0	0.058	0.800E-03	1.373	0.476E-01	24.113	1.228
7.5	0.051	0.577E-03	1.378	0.393E-01	29.216	1.231
8.0	0.045	0.425E-03	1.382	0.328E-01	35.021	1.233
8.5	0.040	0.318E-03	1.386	0.276E-01	41.574	1.235
9.0	0.036	0.241E-03	1.389	0.235E-01	48.923	1.236
9.5	0.032	0.186E-03	1.391	0.201E-01	57.114	1.238
∞	0	0	1.414	0	∞	1.250

TABLE 3 FANNO FLOW OF A PERFECT GAS - ONE-DIMENSIONAL ADIABATIC FLOW OF A PERFECT GAS IN A DUCT OF CONSTANT CROSS-SECTIONAL AREA WITH FRICTION

$\gamma = 1.400$

M	$\frac{T}{T^*}$	$\frac{p}{p^*}$	$\frac{p_o}{p_o^*}$	$\frac{I}{I^*}$	$\frac{4fL_{max}}{D}$
0	1.200	∞	∞	∞	∞
0.02	1.200	54.770	28.942	22.834	1778.451
0.04	1.200	27.382	14.481	11.435	440.352
0.06	1.199	18.251	9.666	7.643	193.031
0.08	1.198	13.684	7.262	5.753	106.718
0.10	1.198	10.944	5.822	4.624	66.922
0.12	1.197	9.116	4.864	3.875	45.408
0.14	1.195	7.809	4.182	3.343	32.511
0.16	1.194	6.829	3.673	2.947	24.198
0.18	1.192	6.066	3.278	2.642	18.543
0.20	1.190	5.455	2.964	2.400	14.533
0.22	1.188	4.955	2.708	2.205	11.596
0.24	1.186	4.538	2.496	2.043	9.386
0.26	1.184	4.185	2.317	1.909	7.688
0.28	1.181	3.882	2.166	1.795	6.357
0.30	1.179	3.619	2.035	1.698	5.299
0.32	1.176	3.389	1.922	1.614	4.447
0.34	1.173	3.185	1.823	1.542	3.752
0.36	1.170	3.004	1.736	1.479	3.180
0.38	1.166	2.842	1.659	1.424	2.705
0.40	1.163	2.696	1.590	1.375	2.308
0.42	1.159	2.563	1.529	1.332	1.974
0.44	1.155	2.443	1.474	1.294	1.692
0.46	1.151	2.333	1.425	1.260	1.451
0.48	1.147	2.231	1.380	1.230	1.245

Table 3

FANNO FLOW OF A PERFECT GAS

$\gamma = 1.400$

M	$\frac{T}{T^*}$	$\frac{p}{p^*}$	$\frac{p_o}{p_o^*}$	$\frac{I}{I^*}$	$\frac{4fL_{max}}{D}$
0.50	1.143	2.138	1.340	1.203	1.069
0.52	1.138	2.052	1.303	1.179	0.917
0.54	1.134	1.972	1.270	1.157	0.787
0.56	1.129	1.898	1.240	1.138	0.674
0.58	1.124	1.828	1.213	1.121	0.576
0.60	1.119	1.763	1.188	1.105	0.491
0.62	1.114	1.703	1.166	1.091	0.417
0.64	1.109	1.646	1.145	1.079	0.353
0.66	1.104	1.592	1.127	1.068	0.298
0.68	1.098	1.541	1.110	1.058	0.250
0.70	1.093	1.493	1.094	1.049	0.208
0.72	1.087	1.448	1.081	1.041	0.172
0.74	1.082	1.405	1.068	1.034	0.141
0.76	1.076	1.365	1.057	1.028	0.114
0.78	1.070	1.326	1.047	1.023	0.092
0.80	1.064	1.289	1.038	1.019	0.072
0.82	1.058	1.254	1.030	1.015	0.056
0.84	1.052	1.221	1.024	1.011	0.042
0.86	1.045	1.189	1.018	1.008	0.031
0.88	1.039	1.158	1.013	1.006	0.022
0.90	1.033	1.129	1.009	1.004	0.015
0.92	1.026	1.101	1.006	1.002	0.009
0.94	1.020	1.074	1.003	1.001	0.005
0.96	1.013	1.049	1.001	1.001	0.002
0.98	1.007	1.024	1.000	1.000	0.000

FANNO FLOW OF A PERFECT GAS

$\gamma = 1.400$

M	$\frac{T}{T^*}$	$\frac{p}{p^*}$	$\frac{p_o}{p_o^*}$	$\frac{I}{I^*}$	$\frac{4fL_{max}}{D}$
1.00	1.000	1.000	1.000	1.000	0.000
1.02	0.993	0.977	1.000	1.000	0.000
1.04	0.987	0.955	1.001	1.001	0.002
1.06	0.980	0.934	1.003	1.001	0.004
1.08	0.973	0.913	1.005	1.002	0.007
1.10	0.966	0.894	1.008	1.003	0.010
1.12	0.959	0.875	1.011	1.004	0.014
1.14	0.952	0.856	1.015	1.006	0.018
1.16	0.946	0.838	1.020	1.007	0.023
1.18	0.939	0.821	1.025	1.009	0.028
1.20	0.932	0.804	1.030	1.011	0.034
1.22	0.925	0.788	1.037	1.013	0.039
1.24	0.918	0.773	1.043	1.015	0.045
1.26	0.911	0.757	1.050	1.017	0.052
1.28	0.904	0.743	1.058	1.019	0.058
1.30	0.897	0.728	1.066	1.022	0.065
1.32	0.890	0.715	1.075	1.024	0.072
1.34	0.883	0.701	1.084	1.027	0.079
1.36	0.876	0.688	1.094	1.029	0.086
1.38	0.869	0.676	1.104	1.032	0.093
1.40	0.862	0.663	1.115	1.035	0.100
1.42	0.855	0.651	1.126	1.037	0.107
1.44	0.848	0.640	1.138	1.040	0.114
1.46	0.841	0.628	1.150	1.043	0.121
1.48	0.834	0.617	1.163	1.046	0.129
1.50	0.828	0.606	1.176	1.049	0.136
1.52	0.821	0.596	1.190	1.052	0.143
1.54	0.814	0.586	1.204	1.055	0.151

FANNO FLOW OF A PERFECT GAS γ = 1.400

M	$\frac{T}{T^*}$	$\frac{p}{p^*}$	$\frac{p_o}{p_o^*}$	$\frac{I}{I^*}$	$\frac{4fL_{max}}{D}$
1.56	0.807	0.576	1.219	1.058	0.158
1.58	0.800	0.566	1.234	1.060	0.165
1.60	0.794	0.557	1.250	1.063	0.172
1.62	0.787	0.548	1.267	1.066	0.180
1.64	0.780	0.539	1.284	1.069	0.187
1.66	0.774	0.530	1.301	1.072	0.194
1.68	0.767	0.521	1.319	1.075	0.201
1.70	0.760	0.513	1.338	1.079	0.208
1.72	0.754	0.505	1.357	1.082	0.215
1.74	0.747	0.497	1.376	1.085	0.222
1.76	0.741	0.489	1.397	1.088	0.228
1.78	0.735	0.481	1.418	1.091	0.235
1.80	0.728	0.474	1.439	1.094	0.242
1.82	0.722	0.467	1.461	1.096	0.249
1.84	0.716	0.460	1.484	1.099	0.255
1.86	0.709	0.453	1.507	1.102	0.262
1.88	0.703	0.446	1.531	1.105	0.268
1.90	0.697	0.439	1.555	1.108	0.274
1.92	0.691	0.433	1.580	1.111	0.281
1.94	0.685	0.427	1.606	1.114	0.287
1.96	0.679	0.420	1.633	1.117	0.293
1.98	0.673	0.414	1.660	1.120	0.299
2.00	0.667	0.408	1.688	1.123	0.305
2.02	0.661	0.402	1.716	1.126	0.311
2.04	0.655	0.397	1.745	1.128	0.317
2.06	0.649	0.391	1.775	1.131	0.323
2.08	0.643	0.386	1.806	1.134	0.328

FANNO FLOW OF A PERFECT GAS

$\gamma = 1.400$

M	$\frac{T}{T^*}$	$\frac{p}{p^*}$	$\frac{p_o}{p_o^*}$	$\frac{I}{I^*}$	$\frac{4fL_{max}}{D}$
2.10	0.638	0.380	1.837	1.137	0.334
2.12	0.632	0.375	1.869	1.139	0.339
2.14	0.626	0.370	1.902	1.142	0.345
2.16	0.621	0.365	1.935	1.145	0.350
2.18	0.615	0.360	1.970	1.147	0.356
2.20	0.610	0.355	2.005	1.150	0.361
2.22	0.604	0.350	2.041	1.153	0.366
2.24	0.599	0.345	2.078	1.155	0.371
2.26	0.594	0.341	2.115	1.158	0.376
2.28	0.588	0.336	2.154	1.160	0.381
2.30	0.583	0.332	2.193	1.163	0.386
2.32	0.578	0.328	2.233	1.165	0.391
2.34	0.563	0.323	2.274	1.168	0.396
2.36	0.568	0.319	2.316	1.170	0.401
2.38	0.563	0.315	2.359	1.173	0.405
2.40	0.558	0.311	2.403	1.175	0.410
2.42	0.553	0.307	2.448	1.177	0.414
2.44	0.548	0.303	2.494	1.180	0.419
2.46	0.543	0.300	2.540	1.182	0.423
2.48	0.538	0.296	2.588	1.184	0.428
2.5	0.533	0.292	2.637	1.187	0.432
2.6	0.510	0.275	2.896	1.198	0.453
2.7	0.488	0.259	3.183	1.208	0.472
2.8	0.467	0.244	3.500	1.218	0.490
2.9	0.447	0.231	3.850	1.228	0.507

FANNO FLOW OF A PERFECT GAS

$\gamma = 1.400$

M	$\frac{T}{T^*}$	$\frac{p}{p^*}$	$\frac{p_o}{p_o^*}$	$\frac{I}{I^*}$	$\frac{4fL_{max}}{D}$
3.0	0.429	0.218	4.235	1.237	0.522
3.1	0.411	0.207	4.657	1.245	0.537
3.2	0.394	0.196	5.121	1.253	0.550
3.3	0.378	0.186	5.629	1.260	0.563
3.4	0.362	0.177	6.184	1.268	0.575
3.5	0.348	0.169	6.790	1.274	0.586
3.6	0.334	0.161	7.450	1.281	0.597
3.7	0.321	0.153	8.169	1.287	0.607
3.8	0.309	0.146	8.951	1.292	0.616
3.9	0.297	0.140	9.799	1.298	0.625
4.0	0.286	0.134	10.7	1.303	0.633
4.5	0.238	0.108	16.6	1.325	0.668
5.0	0.200	0.089	25.0	1.342	0.694
5.5	0.170	0.075	36.9	1.355	0.714
6.0	0.146	0.064	53.2	1.365	0.730
6.5	0.127	0.055	75.1	1.374	0.743
7.0	0.111	0.048	104.1	1.381	0.753
7.5	0.098	0.042	141.8	1.387	0.761
8.0	0.087	0.037	190.1	1.391	0.768
8.5	0.078	0.033	251.1	1.396	0.774
9.0	0.070	0.029	327.2	1.399	0.779
9.5	0.063	0.026	421.1	1.402	0.783
∞	0	0	∞	1.429	0.822

FANNO FLOW OF A PERFECT GAS

$\gamma = 1.333$

M	$\frac{T}{T^*}$	$\frac{p}{p^*}$	$\frac{p_o}{p_o^*}$	$\frac{I}{I^*}$	$\frac{4fL_{max}}{D}$
0	1.167	∞	∞	∞	∞
0.02	1.166	54.001	29.159	23.159	1868.007
0.04	1.166	26.998	14.590	11.597	462.618
0.06	1.166	17.995	9.739	7.750	202.845
0.08	1.165	13.493	7.315	5.833	112.180
0.10	1.165	10.791	5.865	4.687	70.372
0.12	1.164	8.990	4.900	3.927	47.768
0.14	1.163	7.702	4.212	3.388	34.216
0.16	1.162	6.736	3.699	2.986	25.478
0.18	1.160	5.984	3.301	2.676	19.533
0.20	1.159	5.382	2.984	2.430	15.317
0.22	1.157	4.890	2.726	2.231	12.227
0.24	1.155	4.479	2.512	2.067	9.903
0.26	1.154	4.131	2.332	1.930	8.115
0.28	1.151	3.832	2.179	1.814	6.714
0.30	1.149	3.573	2.047	1.715	5.600
0.32	1.147	3.347	1.933	1.630	4.702
0.34	1.144	3.146	1.833	1.557	3.969
0.36	1.142	2.968	1.745	1.492	3.366
0.38	1.139	2.809	1.667	1.436	2.866
0.40	1.136	2.665	1.598	1.386	2.447
0.42	1.133	2.535	1.536	1.342	2.094
0.44	1.130	2.416	1.481	1.303	1.795
0.46	1.127	2.308	1.431	1.268	1.541
0.48	1.123	2.208	1.386	1.237	1.323

Table 3

FANNO FLOW FOR A PERFECT GAS

$\gamma = 1.333$

M	$\frac{T}{T^*}$	$\frac{p}{p^*}$	$\frac{p_o}{p_o^*}$	$\frac{I}{I^*}$	$\frac{4fL_{max}}{D}$
0.50	1.120	2.116	1.345	1.210	1.137
0.52	1.116	2.032	1.308	1.185	0.976
0.54	1.112	1.953	1.275	1.163	0.837
0.56	1.109	1.880	1.244	1.143	0.717
0.58	1.105	1.812	1.217	1.125	0.614
0.60	1.101	1.748	1.192	1.109	0.523
0.62	1.096	1.689	1.169	1.095	0.445
0.64	1.092	1.633	1.148	1.082	0.377
0.66	1.088	1.580	1.129	1.071	0.318
0.68	1.083	1.530	1.112	1.060	0.267
0.70	1.079	1.484	1.096	1.051	0.223
0.72	1.074	1.439	1.082	1.043	0.184
0.74	1.069	1.397	1.070	1.036	0.151
0.76	1.064	1.357	1.058	1.030	0.123
0.78	1.059	1.319	1.048	1.024	0.098
0.80	1.054	1.283	1.039	1.019	0.078
0.82	1.049	1.249	1.031	1.015	0.060
0.84	1.044	1.216	1.024	1.012	0.045
0.86	1.039	1.185	1.018	1.009	0.033
0.88	1.033	1.155	1.013	1.006	0.023
0.90	1.028	1.126	1.009	1.004	0.016
0.92	1.022	1.099	1.006	1.003	0.010
0.94	1.017	1.073	1.003	1.001	0.005
0.96	1.011	1.048	1.001	1.001	0.002
0.98	1.006	1.023	1.000	1.000	0.001

FANNO FLOW OF A PERFECT GAS

$\gamma = 1.333$

M	$\frac{T}{T^*}$	$\frac{p}{p^*}$	$\frac{p_o}{p_o^*}$	$\frac{I}{I^*}$	$\frac{4fL_{max}}{D}$
1.00	1.000	1.000	1.000	1.000	0.000
1.02	0.994	0.978	1.000	1.000	0.000
1.04	0.988	0.956	1.001	1.001	0.002
1.06	0.983	0.935	1.003	1.001	0.004
1.08	0.977	0.915	1.005	1.002	0.007
1.10	0.971	0.896	1.008	1.003	0.011
1.12	0.965	0.877	1.012	1.005	0.015
1.14	0.959	0.859	1.016	1.006	0.020
1.16	0.953	0.842	1.020	1.008	0.025
1.18	0.947	0.825	1.026	1.010	0.031
1.20	0.941	0.808	1.032	1.012	0.037
1.22	0.935	0.793	1.038	1.014	0.043
1.24	0.929	0.777	1.045	1.016	0.049
1.26	0.923	0.762	1.052	1.018	0.056
1.28	0.916	0.748	1.060	1.021	0.063
1.30	0.910	0.734	1.069	1.023	0.071
1.32	0.904	0.720	1.078	1.026	0.078
1.34	0.898	0.707	1.088	1.029	0.086
1.36	0.892	0.694	1.098	1.031	0.093
1.38	0.886	0.682	1.109	1.034	0.101
1.40	0.879	0.670	1.120	1.037	0.109
1.42	0.873	0.658	1.132	1.040	0.117
1.44	0.867	0.647	1.144	1.043	0.125
1.46	0.861	0.636	1.157	1.046	0.133
1.48	0.855	0.625	1.171	1.050	0.141

Table 3

FANNO FLOW OF A PERFECT GAS

$\gamma = 1.333$

M	$\frac{T}{T^*}$	$\frac{p}{p^*}$	$\frac{p_o}{p_o^*}$	$\frac{I}{I^*}$	$\frac{4fL_{max}}{D}$
1.50	0.849	0.614	1.185	1.053	0.149
1.52	0.842	0.604	1.200	1.056	0.157
1.54	0.836	0.594	1.215	1.059	0.165
1.56	0.830	0.584	1.231	1.062	0.173
1.58	0.824	0.575	1.247	1.066	0.181
1.60	0.818	0.565	1.264	1.069	0.190
1.62	0.812	0.556	1.281	1.072	0.198
1.64	0.806	0.547	1.300	1.076	0.205
1.66	0.800	0.539	1.318	1.079	0.213
1.68	0.794	0.530	1.338	1.082	0.221
1.70	0.788	0.522	1.358	1.086	0.229
1.72	0.782	0.514	1.379	1.089	0.237
1.74	0.776	0.506	1.400	1.092	0.245
1.76	0.770	0.498	1.422	1.096	0.252
1.78	0.764	0.491	1.445	1.099	0.260
1.80	0.758	0.484	1.468	1.103	0.267
1.82	0.752	0.476	1.492	1.106	0.275
1.84	0.746	0.469	1.517	1.109	0.282
1.86	0.740	0.463	1.543	1.113	0.289
1.88	0.734	0.456	1.569	1.116	0.297
1.90	0.729	0.449	1.596	1.119	0.304
1.92	0.723	0.443	1.624	1.122	0.311
1.94	0.717	0.437	1.652	1.126	0.318
1.96	0.711	0.430	1.682	1.129	0.325
1.98	0.706	0.424	1.712	1.132	0.332

FANNO FLOW OF A PERFECT GAS

$\gamma = 1.333$

M	$\frac{T}{T^*}$	$\frac{p}{p^*}$	$\frac{p_o}{p_o^*}$	$\frac{I}{I^*}$	$\frac{4fL_{max}}{D}$
2.00	0.700	0.418	1.743	1.136	0.339
2.02	0.695	0.413	1.774	1.139	0.345
2.04	0.689	0.407	1.807	1.142	0.352
2.06	0.684	0.401	1.841	1.145	0.359
2.08	0.678	0.396	1.875	1.148	0.365
2.10	0.673	0.391	1.910	1.151	0.371
2.12	0.667	0.385	1.947	1.155	0.378
2.14	0.662	0.380	1.984	1.158	0.384
2.16	0.657	0.375	2.022	1.161	0.390
2.18	0.651	0.370	2.061	1.164	0.396
2.20	0.646	0.365	2.101	1.167	0.402
2.22	0.641	0.361	2.142	1.170	0.408
2.24	0.636	0.356	2.184	1.173	0.414
2.26	0.630	0.351	2.228	1.176	0.420
2.28	0.625	0.347	2.272	1.179	0.426
2.30	0.620	0.342	2.317	1.182	0.431
2.32	0.615	0.338	2.364	1.185	0.437
2.34	0.610	0.334	2.412	1.187	0.442
2.36	0.605	0.330	2.460	1.190	0.448
2.38	0.600	0.326	2.510	1.193	0.453
2.40	0.595	0.322	2.562	1.196	0.459
2.42	0.591	0.318	2.614	1.199	0.464
2.44	0.586	0.314	2.668	1.201	0.469
2.46	0.581	0.310	2.723	1.204	0.474
2.48	0.576	0.306	2.779	1.207	0.479

FANNO FLOW OF A PERFECT GAS

$\gamma = 1.333$

M	$\frac{T}{T^*}$	$\frac{p}{p^*}$	$\frac{p_o}{p_o^*}$	$\frac{I}{I^*}$	$\frac{4fL_{max}}{D}$
2.5	0.572	0.302	2.837	1.210	0.484
2.6	0.549	0.285	3.147	1.223	0.508
2.7	0.527	0.269	3.494	1.235	0.530
2.8	0.506	0.254	3.883	1.247	0.551
2.9	0.486	0.240	4.319	1.258	0.571
3.0	0.467	0.228	4.805	1.269	0.589
3.1	0.449	0.216	5.346	1.279	0.607
3.2	0.431	0.205	5.949	1.289	0.623
3.3	0.415	0.195	6.618	1.298	0.638
3.4	0.399	0.186	7.361	1.306	0.652
3.5	0.384	0.177	8.184	1.315	0.666
3.6	0.369	0.169	9.094	1.323	0.678
3.7	0.356	0.161	10.100	1.330	0.690
3.8	0.343	0.154	11.210	1.337	0.701
3.9	0.330	0.147	12.432	1.344	0.712
4.0	0.318	0.141	13.8	1.350	0.721
4.5	0.267	0.115	22.7	1.377	0.763
5.0	0.226	0.095	36.6	1.399	0.795
5.5	0.193	0.080	57.6	1.416	0.820
6.0	0.167	0.068	88.4	1.429	0.839
6.5	0.145	0.059	132.7	1.440	0.855
7.0	0.127	0.051	194.9	1.449	0.868
7.5	0.113	0.045	280.7	1.457	0.878
8.0	0.100	0.040	397.0	1.463	0.887
8.5	0.090	0.035	551.9	1.468	0.894
9.0	0.081	0.032	755.6	1.473	0.900
9.5	0.073	0.028	1019.9	1.477	0.905
∞	0	0	∞	1.512	0.953

FANNO FLOW OF A PERFECT GAS

$\gamma = 1.667$

M	$\frac{T}{T^*}$	$\frac{p}{p^*}$	$\frac{p_o}{p_o^*}$	$\frac{I}{I^*}$	$\frac{4fL_{max}}{D}$
0	1.333	∞	∞	∞	∞
0.02	1.333	57.732	28.132	21.662	1493.072
0.04	1.333	28.862	14.077	10.851	369.405
0.06	1.332	19.235	9.397	7.255	161.762
0.08	1.331	14.419	7.061	5.464	89.319
0.10	1.329	11.529	5.662	4.395	55.932
0.12	1.327	9.600	4.732	3.686	37.893
0.14	1.325	8.222	4.070	3.183	27.086
0.16	1.322	7.187	3.576	2.810	20.124
0.18	1.319	6.381	3.193	2.522	15.393
0.20	1.316	5.736	2.888	2.294	12.042
0.22	1.312	5.207	2.640	2.110	9.589
0.24	1.308	4.766	2.435	1.959	7.747
0.26	1.304	4.392	2.262	1.832	6.331
0.28	1.300	4.071	2.115	1.726	5.225
0.30	1.295	3.793	1.989	1.635	4.346
0.32	1.289	3.549	1.880	1.558	3.639
0.34	1.284	3.333	1.784	1.490	3.063
0.36	1.278	3.141	1.700	1.432	2.591
0.38	1.272	2.968	1.626	1.381	2.199
0.40	1.266	2.813	1.560	1.336	1.872
0.42	1.259	2.672	1.501	1.296	1.597
0.44	1.253	2.544	1.449	1.262	1.365
0.46	1.246	2.426	1.401	1.231	1.168
0.48	1.238	2.318	1.359	1.203	1.001

FANNO FLOW OF A PERFECT GAS

$\gamma = 1.667$

M	$\frac{T}{T^*}$	$\frac{p}{p^*}$	$\frac{p_o}{p_o^*}$	$\frac{I}{I^*}$	$\frac{4fL_{max}}{D}$
0.50	1.231	2.219	1.320	1.179	0.857
0.52	1.223	2.127	1.285	1.157	0.734
0.54	1.215	2.042	1.254	1.138	0.627
0.56	1.207	1.962	1.225	1.120	0.536
0.58	1.199	1.888	1.200	1.105	0.457
0.60	1.191	1.819	1.176	1.091	0.389
0.62	1.182	1.754	1.155	1.079	0.330
0.64	1.173	1.692	1.135	1.068	0.278
0.66	1.164	1.635	1.118	1.058	0.234
0.68	1.155	1.581	1.102	1.050	0.196
0.70	1.146	1.529	1.087	1.042	0.163
0.72	1.137	1.481	1.075	1.035	0.134
0.74	1.128	1.435	1.063	1.029	0.110
0.76	1.118	1.391	1.053	1.024	0.089
0.78	1.109	1.350	1.043	1.019	0.071
0.80	1.099	1.310	1.035	1.016	0.056
0.82	1.089	1.273	1.028	1.012	0.043
0.84	1.079	1.237	1.022	1.009	0.033
0.86	1.070	1.203	1.016	1.007	0.024
0.88	1.060	1.170	1.012	1.005	0.017
0.90	1.050	1.138	1.008	1.003	0.011
0.92	1.040	1.108	1.005	1.002	0.007
0.94	1.030	1.080	1.003	1.001	0.004
0.96	1.020	1.052	1.001	1.000	0.002
0.98	1.010	1.025	1.000	1.000	0.000

FANNO FLOW OF A PERFECT GAS

$\gamma = 1.667$

M	$\frac{T}{T^*}$	$\frac{p}{p^*}$	$\frac{p_o}{p_o^*}$	$\frac{I}{I^*}$	$\frac{4fL_{max}}{D}$
1.00	1.000	1.000	1.000	1.000	0.000
1.02	0.990	0.975	1.000	1.000	0.000
1.04	0.980	0.952	1.001	1.000	0.001
1.06	0.970	0.929	1.003	1.001	0.003
1.08	0.960	0.907	1.005	1.002	0.005
1.10	0.950	0.886	1.007	1.002	0.007
1.12	0.940	0.866	1.010	1.003	0.010
1.14	0.930	0.846	1.014	1.005	0.014
1.16	0.920	0.827	1.017	1.006	0.017
1.18	0.911	0.809	1.022	1.007	0.021
1.20	0.901	0.791	1.027	1.008	0.025
1.22	0.891	0.774	1.032	1.010	0.029
1.24	0.881	0.757	1.038	1.012	0.034
1.26	0.872	0.741	1.044	1.013	0.038
1.28	0.862	0.725	1.051	1.015	0.043
1.30	0.853	0.710	1.057	1.017	0.047
1.32	0.843	0.696	1.065	1.019	0.052
1.34	0.834	0.682	1.073	1.020	0.057
1.36	0.825	0.668	1.081	1.022	0.062
1.38	0.816	0.654	1.089	1.024	0.067
1.40	0.806	0.641	1.098	1.026	0.072
1.42	0.797	0.629	1.108	1.028	0.077
1.44	0.788	0.617	1.117	1.030	0.083
1.46	0.779	0.605	1.127	1.032	0.088
1.48	0.771	0.593	1.138	1.034	0.093

FANNO FLOW OF A PERFECT GAS

$\gamma = 1.667$

M	$\frac{T}{T^*}$	$\frac{p}{p^*}$	$\frac{p_o}{p_o^*}$	$\frac{I}{I^*}$	$\frac{4fL_{max}}{D}$
1.50	0.762	0.582	1.148	1.037	0.098
1.52	0.753	0.571	1.160	1.039	0.103
1.54	0.745	0.560	1.171	1.041	0.108
1.56	0.736	0.550	1.183	1.043	0.113
1.58	0.728	0.540	1.195	1.045	0.118
1.60	0.719	0.530	1.208	1.047	0.123
1.62	0.711	0.521	1.220	1.049	0.128
1.64	0.703	0.511	1.234	1.051	0.133
1.66	0.695	0.502	1.247	1.053	0.137
1.68	0.687	0.493	1.261	1.055	0.142
1.70	0.679	0.485	1.275	1.057	0.147
1.72	0.671	0.476	1.290	1.059	0.152
1.74	0.664	0.468	1.305	1.061	0.156
1.76	0.656	0.460	1.320	1.063	0.161
1.78	0.648	0.452	1.336	1.065	0.165
1.80	0.641	0.445	1.352	1.067	0.170
1.82	0.634	0.437	1.368	1.069	0.174
1.84	0.626	0.430	1.385	1.071	0.179
1.86	0.619	0.423	1.402	1.073	0.183
1.88	0.612	0.416	1.419	1.075	0.187
1.90	0.605	0.409	1.437	1.077	0.191
1.92	0.598	0.403	1.455	1.079	0.195
1.94	0.591	0.396	1.474	1.081	0.199
1.96	0.585	0.390	1.492	1.083	0.203
1.98	0.578	0.384	1.512	1.085	0.207

FANNO FLOW OF A PERFECT GAS

$\gamma = 1.667$

M	$\frac{T}{T^*}$	$\frac{p}{p^*}$	$\frac{p_o}{p_o^*}$	$\frac{I}{I^*}$	$\frac{4fL_{max}}{D}$
2.00	0.571	0.378	1.531	1.087	0.211
2.02	0.565	0.372	1.551	1.088	0.215
2.04	0.558	0.366	1.571	1.090	0.219
2.06	0.552	0.361	1.592	1.092	0.223
2.08	0.546	0.355	1.613	1.094	0.226
2.10	0.540	0.350	1.634	1.095	0.230
2.12	0.534	0.345	1.656	1.097	0.233
2.14	0.528	0.339	1.678	1.099	0.237
2.16	0.522	0.334	1.700	1.101	0.240
2.18	0.516	0.329	1.723	1.102	0.244
2.20	0.510	0.325	1.746	1.104	0.247
2.22	0.504	0.320	1.769	1.105	0.250
2.24	0.499	0.315	1.793	1.107	0.254
2.26	0.493	0.311	1.818	1.109	0.257
2.28	0.488	0.306	1.842	1.110	0.260
2.30	0.482	0.302	1.867	1.112	0.263
2.32	0.477	0.298	1.893	1.113	0.266
2.34	0.472	0.294	1.918	1.115	0.269
2.36	0.467	0.289	1.945	1.116	0.272
2.38	0.462	0.285	1.971	1.118	0.275
2.40	0.457	0.282	1.998	1.119	0.278
2.42	0.452	0.278	2.025	1.121	0.280
2.44	0.447	0.274	2.053	1.122	0.283
2.46	0.442	0.270	2.081	1.123	0.286
2.48	0.437	0.267	2.110	1.125	0.289

FANNO FLOW OF A PERFECT GAS

$\gamma = 1.667$

M	$\frac{T}{T^*}$	$\frac{p}{p^*}$	$\frac{p_o}{p_o^*}$	$\frac{I}{I^*}$	$\frac{4fL_{max}}{D}$
2.5	0.432	0.263	2.139	1.126	0.291
2.6	0.410	0.246	2.289	1.133	0.304
2.7	0.389	0.231	2.450	1.139	0.315
2.8	0.369	0.217	2.622	1.144	0.326
2.9	0.350	0.204	2.805	1.150	0.336
3.0	0.333	0.192	2.999	1.155	0.345
3.1	0.317	0.182	3.205	1.159	0.354
3.2	0.302	0.172	3.422	1.164	0.362
3.3	0.288	0.163	3.653	1.168	0.369
3.4	0.275	0.154	3.895	1.172	0.376
3.5	0.262	0.146	4.151	1.175	0.383
3.6	0.251	0.139	4.420	1.178	0.389
3.7	0.240	0.132	4.703	1.182	0.394
3.8	0.229	0.126	5.000	1.185	0.399
3.9	0.220	0.120	5.311	1.187	0.404
4.0	0.210	0.115	5.6	1.190	0.409
4.5	0.172	0.092	7.5	1.201	0.428
5.0	0.143	0.076	9.8	1.209	0.442
5.5	0.120	0.063	12.6	1.216	0.453
6.0	0.103	0.053	15.8	1.221	0.461
6.5	0.088	0.046	19.7	1.225	0.468
7.0	0.077	0.040	24.1	1.228	0.474
7.5	0.067	0.035	29.2	1.231	0.478
8.0	0.060	0.031	35.0	1.233	0.482
8.5	0.053	0.027	41.6	1.235	0.485
9.0	0.048	0.024	48.9	1.236	0.487
9.5	0.043	0.022	57.1	1.238	0.489
∞	0	0	∞	1.250	0.509

TABLE 4 RAYLEIGH FLOW OF A PERFECT GAS - ONE-DIMENSIONAL, FRICTIONLESS FLOW OF A PERFECT GAS IN A DUCT OF CONSTANT CROSS-SECTIONAL AREA WITH STAGNATION TEMPERATURE CHANGE

$\gamma = 1.400$

M	$\frac{T_o}{T_o^*}$	$\frac{T}{T^*}$	$\frac{p}{p^*}$	$\frac{p_o}{p_o^*}$	$\frac{V}{V^*}$
0.00	0.0000	0.0000	2.400	1.268	0.0000
0.02	0.0019	0.0023	2.399	1.268	0.0010
0.04	0.0076	0.0092	2.395	1.266	0.0038
0.06	0.0171	0.0205	2.388	1.265	0.0086
0.08	0.0302	0.0362	2.379	1.262	0.0152
0.10	0.0468	0.0560	2.367	1.259	0.0237
0.12	0.0666	0.0797	2.353	1.255	0.0339
0.14	0.0895	0.1069	2.336	1.251	0.0458
0.16	0.1151	0.1374	2.317	1.246	0.0593
0.18	0.1432	0.1708	2.296	1.241	0.0744
0.20	0.1736	0.2066	2.273	1.235	0.0909
0.22	0.2057	0.2445	2.248	1.228	0.1088
0.24	0.2395	0.2841	2.221	1.221	0.1279
0.26	0.2745	0.3250	2.193	1.214	0.1482
0.28	0.3104	0.3667	2.163	1.206	0.1696
0.30	0.3469	0.4089	2.131	1.199	0.1918
0.32	0.3837	0.4512	2.099	1.190	0.2149
0.34	0.4206	0.4933	2.066	1.182	0.2388
0.36	0.4572	0.5348	2.031	1.174	0.2633
0.38	0.4935	0.5755	1.996	1.165	0.2883
0.40	0.5290	0.6151	1.961	1.157	0.3137
0.42	0.5638	0.6535	1.925	1.148	0.3395
0.44	0.5975	0.6903	1.888	1.139	0.3656
0.46	0.6301	0.7254	1.852	1.131	0.3918
0.48	0.6614	0.7587	1.815	1.122	0.4181

RAYLEIGH FLOW OF A PERFECT GAS

$\gamma = 1.400$

M	$\frac{T_o}{T_o^*}$	$\frac{T}{T^*}$	$\frac{p}{p^*}$	$\frac{p_o}{p_o^*}$	$\frac{V}{V^*}$
0.50	0.6914	0.7901	1.778	1.114	0.4444
0.52	0.7199	0.8196	1.741	1.106	0.4708
0.54	0.7470	0.8469	1.704	1.098	0.4970
0.56	0.7725	0.8723	1.668	1.090	0.5230
0.58	0.7965	0.8955	1.632	1.083	0.5489
0.60	0.8189	0.9167	1.596	1.075	0.5745
0.62	0.8398	0.9358	1.560	1.068	0.5998
0.64	0.8592	0.9530	1.525	1.061	0.6248
0.66	0.8771	0.9682	1.491	1.055	0.6494
0.68	0.8935	0.9814	1.457	1.049	0.6737
0.70	0.9085	0.9929	1.423	1.043	0.6975
0.72	0.9221	1.0026	1.391	1.038	0.7209
0.74	0.9344	1.0106	1.359	1.033	0.7439
0.76	0.9455	1.0171	1.327	1.028	0.7665
0.78	0.9553	1.0220	1.296	1.023	0.7885
0.80	0.9639	1.0255	1.266	1.019	0.8101
0.82	0.9715	1.0276	1.236	1.016	0.8313
0.84	0.9781	1.0285	1.207	1.012	0.8519
0.86	0.9836	1.0283	1.179	1.010	0.8721
0.88	0.9883	1.0269	1.152	1.007	0.8918
0.90	0.9921	1.0245	1.125	1.005	0.9110
0.92	0.9951	1.0212	1.098	1.003	0.9297
0.94	0.9973	1.0170	1.073	1.002	0.9480
0.96	0.9988	1.0120	1.048	1.001	0.9658
0.98	0.9997	1.0064	1.024	1.000	0.9831

RAYLEIGH FLOW OF A PERFECT GAS

$\gamma = 1.400$

M	$\frac{T_o}{T_o^*}$	$\frac{T}{T^*}$	$\frac{p}{p^*}$	$\frac{p_o}{p_o^*}$	$\frac{V}{V^*}$
1.00	1.000	1.000	1.000	1.000	1.000
1.02	1.000	0.993	0.977	1.000	1.016
1.04	0.999	0.986	0.955	1.001	1.032
1.06	0.998	0.978	0.933	1.002	1.048
1.08	0.996	0.969	0.912	1.003	1.063
1.10	0.994	0.960	0.891	1.005	1.078
1.12	0.991	0.951	0.871	1.007	1.092
1.14	0.989	0.942	0.851	1.010	1.106
1.16	0.986	0.932	0.832	1.012	1.120
1.18	0.982	0.922	0.814	1.016	1.133
1.20	0.979	0.912	0.796	1.019	1.146
1.22	0.975	0.902	0.778	1.023	1.158
1.24	0.971	0.891	0.761	1.028	1.171
1.26	0.967	0.881	0.745	1.033	1.182
1.28	0.962	0.870	0.729	1.038	1.194
1.30	0.958	0.859	0.713	1.044	1.205
1.32	0.953	0.848	0.698	1.050	1.216
1.34	0.949	0.838	0.683	1.056	1.226
1.36	0.944	0.827	0.669	1.063	1.237
1.38	0.939	0.816	0.655	1.070	1.247
1.40	0.934	0.805	0.641	1.078	1.256
1.42	0.929	0.795	0.628	1.086	1.266
1.44	0.924	0.784	0.615	1.094	1.275
1.46	0.919	0.773	0.602	1.103	1.284
1.48	0.914	0.763	0.590	1.112	1.293

RAYLEIGH FLOW OF A PERFECT GAS

$\gamma = 1.400$

M	$\frac{T_o}{T_o^*}$	$\frac{T}{T^*}$	$\frac{p}{p^*}$	$\frac{p_o}{p_o^*}$	$\frac{V}{V^*}$
1.50	0.909	0.753	0.578	1.122	1.301
1.52	0.904	0.742	0.567	1.132	1.309
1.54	0.899	0.732	0.556	1.142	1.317
1.56	0.894	0.722	0.545	1.153	1.325
1.58	0.889	0.712	0.534	1.164	1.333
1.60	0.884	0.702	0.524	1.176	1.340
1.62	0.879	0.692	0.513	1.188	1.348
1.64	0.874	0.682	0.504	1.200	1.355
1.66	0.869	0.673	0.494	1.213	1.361
1.68	0.865	0.663	0.485	1.226	1.368
1.70	0.860	0.654	0.476	1.240	1.375
1.72	0.855	0.645	0.467	1.254	1.381
1.74	0.850	0.635	0.458	1.269	1.387
1.76	0.846	0.626	0.450	1.284	1.393
1.78	0.841	0.618	0.442	1.300	1.399
1.80	0.836	0.609	0.434	1.316	1.405
1.82	0.832	0.600	0.426	1.332	1.410
1.84	0.827	0.592	0.418	1.349	1.416
1.86	0.823	0.584	0.411	1.367	1.421
1.88	0.818	0.575	0.403	1.385	1.426
1.90	0.814	0.567	0.396	1.403	1.431
1.92	0.810	0.559	0.390	1.422	1.436
1.94	0.806	0.552	0.383	1.442	1.441
1.96	0.802	0.544	0.376	1.462	1.446
1.98	0.797	0.536	0.370	1.482	1.450

RAYLEIGH FLOW OF A PERFECT GAS

$\gamma = 1.400$

M	$\frac{To}{To^*}$	$\frac{T}{T^*}$	$\frac{p}{p^*}$	$\frac{p_o}{p_o^*}$	$\frac{V}{V^*}$
2.00	0.793	0.529	0.364	1.503	1.455
2.02	0.789	0.522	0.358	1.525	1.459
2.04	0.785	0.514	0.352	1.547	1.463
2.06	0.782	0.507	0.346	1.569	1.467
2.08	0.778	0.500	0.340	1.592	1.471
2.10	0.774	0.494	0.335	1.616	1.475
2.12	0.770	0.487	0.329	1.640	1.479
2.14	0.767	0.480	0.324	1.665	1.483
2.16	0.763	0.474	0.319	1.691	1.487
2.18	0.760	0.467	0.314	1.717	1.490
2.20	0.756	0.461	0.309	1.743	1.494
2.22	0.753	0.455	0.304	1.771	1.497
2.24	0.749	0.449	0.299	1.799	1.501
2.26	0.746	0.443	0.294	1.827	1.504
2.28	0.743	0.437	0.290	1.856	1.507
2.30	0.740	0.431	0.286	1.886	1.510
2.32	0.736	0.426	0.281	1.916	1.513
2.34	0.733	0.420	0.277	1.948	1.516
2.36	0.730	0.415	0.273	1.979	1.519
2.38	0.727	0.409	0.269	2.012	1.522
2.40	0.724	0.404	0.265	2.045	1.525
2.42	0.721	0.399	0.261	2.079	1.528
2.44	0.718	0.394	0.257	2.114	1.531
2.46	0.716	0.388	0.253	2.149	1.533
2.48	0.713	0.384	0.250	2.185	1.536

RAYLEIGH FLOW OF A PERFECT GAS

$\gamma = 1.400$

M	$\frac{T_o}{T_o^*}$	$\frac{T}{T^*}$	$\frac{p}{p^*}$	$\frac{P_o}{P_o^*}$	$\frac{V}{V^*}$
2.5	0.710	0.379	0.246	2.222	1.538
2.6	0.697	0.356	0.229	2.418	1.550
2.7	0.685	0.334	0.214	2.634	1.561
2.8	0.674	0.315	0.200	2.873	1.571
2.9	0.664	0.297	0.188	3.136	1.580
3.0	0.654	0.280	0.176	3.424	1.588
3.1	0.645	0.265	0.166	3.741	1.596
3.2	0.637	0.251	0.156	4.087	1.603
3.3	0.629	0.238	0.148	4.465	1.609
3.4	0.622	0.225	0.140	4.878	1.615
3.5	0.616	0.214	0.132	5.328	1.620
3.6	0.610	0.204	0.125	5.817	1.625
3.7	0.604	0.194	0.119	6.349	1.629
3.8	0.599	0.185	0.113	6.926	1.633
3.9	0.594	0.176	0.108	7.550	1.637
4.0	0.589	0.1683	0.1026	8.23	1.641
4.5	0.570	0.1354	0.0818	12.50	1.656
5.0	0.556	0.1111	0.0667	18.63	1.667
5.5	0.545	0.0927	0.0554	27.21	1.675
6.0	0.536	0.0785	0.0467	38.95	1.681
6.5	0.530	0.0673	0.0399	54.68	1.686
7.0	0.524	0.0583	0.0345	75.41	1.690
7.5	0.520	0.0509	0.0301	102.29	1.693
8.0	0.516	0.0449	0.0265	136.62	1.695
8.5	0.513	0.0399	0.0235	179.92	1.698
9.0	0.511	0.0356	0.0210	233.88	1.699
9.5	0.509	0.0321	0.0188	300.41	1.701
∞	0.490	0	0	∞	1.714

RAYLEIGH FLOW OF A PERFECT GAS

$\gamma = 1.333$

M	$\frac{T_o}{T_o*}$	$\frac{T}{T*}$	$\frac{p}{p*}$	$\frac{p_o}{p_o*}$	$\frac{V}{V*}$
0.00	0.0000	0.0000	2.333	1.259	0.0000
0.02	0.0019	0.0022	2.332	1.259	0.0009
0.04	0.0074	0.0087	2.328	1.258	0.0037
0.06	0.0166	0.0194	2.322	1.256	0.0084
0.08	0.0294	0.0342	2.313	1.254	0.0148
0.10	0.0455	0.0530	2.302	1.251	0.0230
0.12	0.0648	0.0755	2.289	1.248	0.0330
0.14	0.0871	0.1013	2.274	1.243	0.0446
0.16	0.1122	0.1303	2.256	1.239	0.0578
0.18	0.1397	0.1620	2.236	1.234	0.0725
0.20	0.1693	0.1962	2.215	1.228	0.0886
0.22	0.2009	0.2325	2.192	1.222	0.1061
0.24	0.2340	0.2704	2.167	1.215	0.1248
0.26	0.2684	0.3096	2.140	1.208	0.1447
0.28	0.3038	0.3498	2.112	1.201	0.1656
0.30	0.3398	0.3905	2.083	1.194	0.1875
0.32	0.3762	0.4315	2.053	1.186	0.2102
0.34	0.4128	0.4724	2.021	1.178	0.2337
0.36	0.4492	0.5129	1.989	1.170	0.2578
0.38	0.4852	0.5527	1.956	1.162	0.2825
0.40	0.5207	0.5916	1.923	1.153	0.3077
0.42	0.5554	0.6294	1.889	1.145	0.3332
0.44	0.5891	0.6658	1.854	1.137	0.3590
0.46	0.6218	0.7007	1.820	1.128	0.3851
0.48	0.6533	0.7340	1.785	1.120	0.4112

RAYLEIGH FLOW OF A PERFECT GAS

$\gamma = 1.333$

M	$\frac{To}{To^*}$	$\frac{T}{T^*}$	$\frac{p}{p^*}$	$\frac{p_o}{p_o^*}$	$\frac{V}{V^*}$
0.50	0.6836	0.7655	1.750	1.112	0.4375
0.52	0.7124	0.7952	1.715	1.104	0.4637
0.54	0.7398	0.8230	1.680	1.096	0.4899
0.56	0.7657	0.8489	1.645	1.089	0.5159
0.58	0.7901	0.8728	1.611	1.081	0.5418
0.60	0.8130	0.8947	1.576	1.074	0.5675
0.62	0.8343	0.9147	1.543	1.067	0.5930
0.64	0.8542	0.9328	1.509	1.061	0.6181
0.66	0.8725	0.9490	1.476	1.055	0.6429
0.68	0.8894	0.9633	1.443	1.049	0.6674
0.70	0.9048	0.9759	1.411	1.043	0.6915
0.72	0.9189	0.9867	1.380	1.037	0.7152
0.74	0.9316	0.9959	1.349	1.032	0.7385
0.76	0.9430	1.0035	1.318	1.028	0.7613
0.78	0.9532	1.0097	1.288	1.023	0.7838
0.80	0.9623	1.0144	1.259	1.019	0.8057
0.82	0.9702	1.0177	1.230	1.016	0.8272
0.84	0.9770	1.0198	1.202	1.012	0.8483
0.86	0.9828	1.0207	1.175	1.010	0.8689
0.88	0.9877	1.0205	1.148	1.007	0.8890
0.90	0.9917	1.0193	1.122	1.005	0.9086
0.92	0.9948	1.0171	1.096	1.003	0.9278
0.94	0.9971	1.0140	1.071	1.002	0.9466
0.96	0.9988	1.0101	1.047	1.001	0.9648
0.98	0.9997	1.0054	1.023	1.000	0.9826

RAYLEIGH FLOW OF A PERFECT GAS

$\gamma = 1.333$

M	$\frac{To}{To*}$	$\frac{T}{T*}$	$\frac{p}{p*}$	$\frac{p_o}{p_o*}$	$\frac{V}{V*}$
1.00	1.000	1.000	1.000	1.000	1.000
1.02	1.000	0.994	0.977	1.000	1.017
1.04	0.999	0.987	0.955	1.001	1.033
1.06	0.998	0.980	0.934	1.002	1.049
1.08	0.996	0.973	0.913	1.003	1.065
1.10	0.994	0.965	0.893	1.005	1.080
1.12	0.991	0.956	0.873	1.007	1.095
1.14	0.988	0.947	0.854	1.010	1.110
1.16	0.985	0.938	0.835	1.013	1.124
1.18	0.981	0.929	0.817	1.016	1.137
1.20	0.977	0.920	0.799	1.020	1.151
1.22	0.973	0.910	0.782	1.024	1.164
1.24	0.969	0.900	0.765	1.028	1.176
1.26	0.964	0.890	0.749	1.033	1.189
1.28	0.960	0.880	0.733	1.039	1.201
1.30	0.955	0.869	0.717	1.045	1.212
1.32	0.950	0.859	0.702	1.051	1.223
1.34	0.945	0.849	0.687	1.057	1.234
1.36	0.940	0.838	0.673	1.064	1.245
1.38	0.935	0.828	0.659	1.072	1.256
1.40	0.929	0.817	0.646	1.080	1.266
1.42	0.924	0.807	0.633	1.088	1.276
1.44	0.919	0.797	0.620	1.097	1.285
1.46	0.913	0.786	0.607	1.106	1.295
1.48	0.908	0.776	0.595	1.115	1.304

RAYLEIGH FLOW OF A PERFECT GAS

$\gamma = 1.333$

M	$\frac{T_o}{T_o^*}$	$\frac{T}{T^*}$	$\frac{p}{p^*}$	$\frac{p_o}{p_o^*}$	$\frac{V}{V^*}$
1.50	0.902	0.766	0.583	1.126	1.313
1.52	0.897	0.756	0.572	1.136	1.321
1.54	0.891	0.745	0.561	1.147	1.330
1.56	0.886	0.735	0.550	1.158	1.338
1.58	0.880	0.725	0.539	1.170	1.346
1.60	0.875	0.716	0.529	1.182	1.354
1.62	0.870	0.706	0.519	1.195	1.361
1.64	0.864	0.696	0.509	1.208	1.368
1.66	0.859	0.687	0.499	1.222	1.376
1.68	0.854	0.677	0.490	1.236	1.383
1.70	0.848	0.668	0.481	1.251	1.389
1.72	0.843	0.659	0.472	1.266	1.396
1.74	0.838	0.650	0.463	1.282	1.403
1.76	0.833	0.641	0.455	1.298	1.409
1.78	0.828	0.632	0.447	1.314	1.415
1.80	0.823	0.623	0.439	1.332	1.421
1.82	0.818	0.615	0.431	1.349	1.427
1.84	0.813	0.606	0.423	1.368	1.433
1.86	0.808	0.598	0.416	1.387	1.438
1.88	0.803	0.590	0.408	1.406	1.444
1.90	0.798	0.582	0.401	1.426	1.449
1.92	0.794	0.574	0.394	1.446	1.454
1.94	0.789	0.566	0.388	1.468	1.459
1.96	0.784	0.558	0.381	1.489	1.464
1.98	0.780	0.550	0.375	1.512	1.469

RAYLEIGH FLOW OF A PERFECT GAS

$\gamma = 1.333$

M	$\frac{T_o}{T_o^*}$	$\frac{T}{T^*}$	$\frac{p}{p^*}$	$\frac{p_o}{p_o^*}$	$\frac{V}{V^*}$
2.00	0.776	0.543	0.368	1.535	1.474
2.02	0.771	0.536	0.362	1.558	1.478
2.04	0.767	0.528	0.356	1.582	1.483
2.06	0.763	0.521	0.350	1.607	1.487
2.08	0.758	0.514	0.345	1.633	1.492
2.10	0.754	0.507	0.339	1.659	1.496
2.12	0.750	0.501	0.334	1.686	1.500
2.14	0.746	0.494	0.328	1.714	1.504
2.16	0.742	0.487	0.323	1.742	1.508
2.18	0.738	0.481	0.318	1.771	1.512
2.20	0.734	0.474	0.313	1.801	1.515
2.22	0.731	0.468	0.308	1.831	1.519
2.24	0.727	0.462	0.303	1.862	1.523
2.26	0.723	0.456	0.299	1.894	1.526
2.28	0.720	0.450	0.294	1.927	1.529
2.30	0.716	0.444	0.290	1.961	1.533
2.32	0.713	0.438	0.285	1.995	1.536
2.34	0.709	0.433	0.281	2.031	1.539
2.36	0.706	0.427	0.277	2.067	1.542
2.38	0.702	0.422	0.273	2.104	1.546
2.40	0.699	0.416	0.269	2.142	1.549
2.42	0.696	0.411	0.265	2.181	1.551
2.44	0.693	0.406	0.261	2.220	1.554
2.46	0.690	0.401	0.257	2.261	1.557
2.48	0.686	0.396	0.254	2.303	1.560

RAYLEIGH FLOW OF A PERFECT GAS

$\gamma = 1.333$

M	$\frac{To}{To^*}$	$\frac{T}{T^*}$	$\frac{p}{p^*}$	$\frac{p_o}{p_o^*}$	$\frac{V}{V^*}$
2.5	0.683	0.391	0.250	2.345	1.563
2.6	0.669	0.367	0.233	2.574	1.575
2.7	0.656	0.345	0.218	2.829	1.587
2.8	0.643	0.325	0.204	3.114	1.597
2.9	0.632	0.307	0.191	3.433	1.607
3.0	0.621	0.290	0.180	3.787	1.616
3.1	0.611	0.274	0.169	4.180	1.623
3.2	0.602	0.260	0.159	4.616	1.631
3.3	0.594	0.246	0.150	5.099	1.637
3.4	0.586	0.234	0.142	5.634	1.644
3.5	0.579	0.222	0.135	6.225	1.649
3.6	0.572	0.211	0.128	6.876	1.654
3.7	0.565	0.201	0.121	7.594	1.659
3.8	0.559	0.192	0.115	8.384	1.664
3.9	0.554	0.183	0.110	9.253	1.668
4.0	0.549	0.1747	0.1045	10.21	1.672
4.5	0.527	0.1407	0.0833	16.51	1.688
5.0	0.511	0.1155	0.0680	26.19	1.699
5.5	0.499	0.0964	0.0565	40.69	1.708
6.0	0.490	0.0816	0.0476	61.88	1.714
6.5	0.482	0.0700	0.0407	92.14	1.720
7.0	0.476	0.0606	0.0352	134.50	1.724
7.5	0.471	0.0530	0.0307	192.71	1.727
8.0	0.467	0.0468	0.0270	271.33	1.730
8.5	0.464	0.0415	0.0240	375.92	1.732
9.0	0.461	0.0371	0.0214	513.09	1.734
9.5	0.459	0.0334	0.0192	690.70	1.736
∞	0.437	0	0	∞	1.750

RAYLEIGH FLOW OF A PERFECT GAS

$\gamma = 1.667$

M	$\frac{To}{To^*}$	$\frac{T}{T^*}$	$\frac{p}{p^*}$	$\frac{p_o}{p_o^*}$	$\frac{V}{V^*}$
0.00	0.0000	0.0000	2.667	1.299	0.0000
0.02	0.0021	0.0028	2.665	1.299	0.0011
0.04	0.0085	0.0113	2.660	1.297	0.0043
0.06	0.0190	0.0253	2.651	1.295	0.0095
0.08	0.0335	0.0446	2.639	1.292	0.0169
0.10	0.0518	0.0688	2.623	1.288	0.0262
0.12	0.0736	0.0977	2.604	1.284	0.0375
0.14	0.0987	0.1307	2.583	1.279	0.0506
0.16	0.1267	0.1675	2.558	1.273	0.0655
0.18	0.1572	0.2074	2.530	1.266	0.0820
0.20	0.1900	0.2501	2.500	1.259	0.1000
0.22	0.2245	0.2948	2.468	1.251	0.1194
0.24	0.2607	0.3411	2.433	1.243	0.1402
0.26	0.2978	0.3884	2.397	1.234	0.1620
0.28	0.3357	0.4362	2.359	1.225	0.1849
0.30	0.3739	0.4840	2.319	1.216	0.2087
0.32	0.4121	0.5314	2.278	1.207	0.2333
0.34	0.4502	0.5780	2.236	1.197	0.2585
0.36	0.4877	0.6234	2.193	1.187	0.2842
0.38	0.5244	0.6672	2.150	1.178	0.3104
0.40	0.5603	0.7093	2.105	1.168	0.3369
0.42	0.5949	0.7493	2.061	1.158	0.3636
0.44	0.6283	0.7871	2.016	1.148	0.3904
0.46	0.6603	0.8225	1.972	1.139	0.4172
0.48	0.6908	0.8555	1.927	1.129	0.4440

Table 4

RAYLEIGH FLOW OF A PERFECT GAS

$\gamma = 1.667$

M	$\frac{T_o}{T_o^*}$	$\frac{T}{T^*}$	$\frac{p}{p^*}$	$\frac{p_o}{p_o^*}$	$\frac{V}{V^*}$
0.50	0.7198	0.8859	1.882	1.120	0.4706
0.52	0.7471	0.9138	1.838	1.111	0.4971
0.54	0.7728	0.9392	1.795	1.102	0.5233
0.56	0.7968	0.9619	1.751	1.094	0.5492
0.58	0.8192	0.9822	1.709	1.086	0.5748
0.60	0.8400	1.0001	1.667	1.078	0.6000
0.62	0.8592	1.0156	1.625	1.070	0.6248
0.64	0.8769	1.0288	1.585	1.063	0.6492
0.66	0.8931	1.0399	1.545	1.056	0.6730
0.68	0.9078	1.0489	1.506	1.050	0.6964
0.70	0.9219	1.0559	1.468	1.044	0.7193
0.72	0.9333	1.0611	1.431	1.038	0.7417
0.74	0.9441	1.0645	1.394	1.033	0.7635
0.76	0.9537	1.0663	1.359	1.028	0.7848
0.78	0.9622	1.0667	1.324	1.023	0.8056
0.80	0.9697	1.0656	1.290	1.019	0.8258
0.82	0.9761	1.0633	1.257	1.016	0.8455
0.84	0.9817	1.0597	1.226	1.012	0.8647
0.86	0.9864	1.0551	1.194	1.009	0.8834
0.88	0.9903	1.0495	1.164	1.007	0.9015
0.90	0.9935	1.0430	1.135	1.005	0.9192
0.92	0.9959	1.0357	1.106	1.003	0.9363
0.94	0.9978	1.0277	1.078	1.002	0.9529
0.96	0.9990	1.0190	1.052	1.001	0.9691
0.98	0.9998	1.0098	1.025	1.000	0.9848

RAYLEIGH FLOW OF A PERFECT GAS

$\gamma = 1.667$

M	$\frac{T_o}{T_o^*}$	$\frac{T}{T^*}$	$\frac{p}{p^*}$	$\frac{p_o}{p_o^*}$	$\frac{V}{V^*}$
1.00	1.000	1.000	1.000	1.000	1.000
1.02	1.000	0.990	0.975	1.000	1.015
1.04	0.999	0.979	0.951	1.001	1.029
1.06	0.998	0.968	0.928	1.002	1.043
1.08	0.997	0.957	0.906	1.003	1.057
1.10	0.995	0.945	0.884	1.005	1.070
1.12	0.993	0.934	0.863	1.007	1.082
1.14	0.991	0.922	0.842	1.009	1.095
1.16	0.989	0.910	0.822	1.012	1.107
1.18	0.986	0.898	0.803	1.015	1.118
1.20	0.983	0.886	0.784	1.018	1.129
1.22	0.980	0.874	0.766	1.022	1.140
1.24	0.977	0.861	0.748	1.026	1.151
1.26	0.974	0.849	0.731	1.030	1.161
1.28	0.971	0.837	0.715	1.035	1.171
1.30	0.967	0.825	0.699	1.040	1.181
1.32	0.964	0.813	0.683	1.045	1.190
1.34	0.960	0.801	0.668	1.051	1.199
1.36	0.957	0.789	0.653	1.057	1.208
1.38	0.953	0.777	0.639	1.063	1.217
1.40	0.949	0.766	0.625	0.070	1.225
1.42	0.946	0.754	0.612	1.077	1.233
1.44	0.942	0.743	0.598	1.084	1.241
1.46	0.938	0.731	0.586	1.092	1.249
1.48	0.935	0.720	0.573	1.100	1.256

RAYLEIGH FLOW OF A PERFECT GAS

$\gamma = 1.667$

M	$\frac{T_o}{T_o^*}$	$\frac{T}{T^*}$	$\frac{p}{p^*}$	$\frac{p_o}{p_o^*}$	$\frac{V}{V^*}$
1.50	0.931	0.709	0.561	1.108	1.263
1.52	0.927	0.698	0.550	1.116	1.270
1.54	0.923	0.687	0.538	1.125	1.277
1.56	0.920	0.677	0.527	1.134	1.283
1.58	0.916	0.667	0.517	1.144	1.290
1.60	0.912	0.656	0.506	1.153	1.296
1.62	0.909	0.646	0.496	1.163	1.302
1.64	0.905	0.636	0.486	1.174	1.308
1.66	0.901	0.626	0.477	1.184	1.314
1.68	0.898	0.617	0.467	1.195	1.319
1.70	0.894	0.607	0.458	1.206	1.325
1.72	0.891	0.598	0.450	1.218	1.330
1.74	0.888	0.589	0.441	1.229	1.335
1.76	0.884	0.580	0.433	1.241	1.340
1.78	0.881	0.571	0.425	1.254	1.345
1.80	0.878	0.562	0.417	1.266	1.350
1.82	0.874	0.554	0.409	1.279	1.355
1.84	0.871	0.545	0.401	1.293	1.359
1.86	0.868	0.537	0.394	1.306	1.363
1.88	0.865	0.529	0.387	1.320	1.368
1.90	0.862	0.521	0.380	1.334	1.372
1.92	0.859	0.514	0.373	o.348	1.376
1.94	0.856	0.506	0.367	1.363	1.380
1.96	0.853	0.498	0.360	1.378	1.384
1.98	0.850	0.491	0.354	1.393	1.388

RAYLEIGH FLOW OF A PERFECT GAS

$\gamma = 1.667$

M	$\frac{T_o}{T_o^*}$	$\frac{T}{T^*}$	$\frac{p}{p^*}$	$\frac{p_o}{p_o^*}$	$\frac{V}{V^*}$
2.00	0.847	0.484	0.348	1.409	1.391
2.02	0.844	0.477	0.342	1.425	1.395
2.04	0.841	0.470	0.336	1.441	1.398
2.06	0.839	0.463	0.330	1.458	1.402
2.08	0.836	0.456	0.325	1.474	1.405
2.10	0.833	0.450	0.319	1.492	1.408
2.12	0.831	0.443	0.314	1.509	1.411
2.14	0.828	0.437	0.309	1.527	1.415
2.16	0.826	0.431	0.304	1.545	1.418
2.18	0.823	0.425	0.299	1.563	1.421
2.20	0.821	0.419	0.294	1.582	1.423
2.22	0.818	0.413	0.289	1.601	1.426
2.24	0.816	0.407	0.285	1.620	1.429
2.26	0.814	0.401	0.280	1.639	1.432
2.28	0.811	0.396	0.276	1.659	1.434
2.30	0.809	0.390	0.272	1.680	1.437
2.32	0.807	0.385	0.267	1.700	1.439
2.34	0.805	0.380	0.263	1.721	1.442
2.36	0.803	0.375	0.259	1.742	1.444
2.38	0.800	0.369	0.255	1.764	1.447
2.40	0.798	0.365	0.252	1.785	1.449
2.42	0.796	0.360	0.248	1.807	1.451
2.44	0.794	0.355	0.244	1.830	1.453
2.46	0.792	0.350	0.241	1.853	1.456
2.48	0.791	0.345	0.237	1.876	1.458

Table 4

RAYLEIGH FLOW OF A PERFECT GAS

$\gamma = 1.667$

M	$\frac{T_o}{T_o^*}$	$\frac{T}{T^*}$	$\frac{p}{p^*}$	$\frac{p_o}{p_o^*}$	$\frac{V}{V^*}$
2.5	0.789	0.341	0.234	1.899	1.460
2.6	0.780	0.319	0.217	2.021	1.469
2.7	0.771	0.300	0.203	2.152	1.478
2.8	0.764	0.282	0.190	2.291	1.486
2.9	0.757	0.265	0.178	2.440	1.493
3.0	0.750	0.250	0.167	2.597	1.500
3.1	0.744	0.236	0.157	2.764	1.506
3.2	0.739	0.223	0.148	2.941	1.511
3.3	0.733	0.211	0.139	3.128	1.516
3.4	0.729	0.200	0.132	3.325	1.521
3.5	0.724	0.190	0.125	3.532	1.525
3.6	0.720	0.180	0.118	3.751	1.529
3.7	0.716	0.172	0.112	3.980	1.533
3.8	0.713	0.163	0.106	4.221	1.536
3.9	0.709	0.156	0.101	4.473	1.539
4.0	0.706	0.1486	0.0964	4.74	1.542
4.5	0.693	0.1192	0.0767	6.25	1.554
5.0	0.684	0.0976	0.0625	8.10	1.562
5.5	0.677	0.0814	0.0519	10.32	1.569
6.0	0.671	0.0688	0.0437	12.96	1.574
6.5	0.667	0.0589	0.0373	16.06	1.577
7.0	0.663	0.0510	0.0323	19.63	1.581
7.5	0.660	0.0445	0.0281	23.74	1.583
8.0	0.658	0.0393	0.0248	28.40	1.585
8.5	0.656	0.0348	0.0220	33.67	1.587
9.0	0.654	0.0311	0.0196	39.57	1.588
9.5	0.653	0.0280	0.0176	46.14	1.589
∞	0.640	0	0	∞	1.600

TABLE 5 THE STANDARD ATMOSPHERE

The following average properties of the atmosphere in temperate latitudes have been standardized by international agreement for use in aeronautical calculations.

Detailed properties for altitudes up to 20,000 metres are given in NACA Tech. Note 3182.

At sea level the properties are:

Temperature (T_{sl})	288.15 K $15^{o}C$
Pressure (P_{sl})	101.325 kN/m^2 760 mm Hg
Density (ρ_{sl})	1.2254 kg/m^3
Kinematic viscosity (ν_{sl})	10^{-5} x 1.4638 m^2/s
Speed of sound (a_{sl})	340.429 m/s

Temperature, pressure, density, and kinematic viscosity at altitude are tabulated on p. 173. The speed of sound at altitude is given by

$$a/a_{sl} = (T/T_{sl})^{\frac{1}{2}}.$$

<u>Properties of the standard atmosphere at altitude</u>

Height (metres)	Temperature ratio (absolute temperatures) (T/T_{sl})	Pressure ratio (p/p_{sl})	Density ratio (ρ/ρ_{sl})	Kinematic viscosity-ratio (ν/ν_{sl})
0	1.0000	1.0000	1.0000	1.0000
1,000	0.977	0.887	0.907	1.082
2,000	0.955	0.784	0.822	1.173
3,000	0.932	0.692	0.742	1.274
4,000	0.910	0.608	0.669	1.386
5,000	0.887	0.533	0.601	1.511
6,000	0.865	0.466	0.538	1.651
7,000	0.842	0.405	0.481	1.807
8,000	0.819	0.351	0.429	1.983
9,000	0.797	0.303	0.381	2.182
10,000	0.774	0.261	0.337	2.406
11,000	0.752	0.223	0.297	2.661
12,000	0.752	0.191	0.254	3.115
13,000	0.752	0.163	0.217	3.647
14,000	0.752	0.139	0.185	4.270
15,000	0.752	0.119	0.158	5.000
16,000	0.752	0.101	0.135	5.854
17,000	0.752	0.087	0.115	6.853
18,000	0.752	0.074	0.098	8.024
19,000	0.752	0.063	0.084	9.395
20,000	0.752	0.054	0.072	10.999
21.000	0.752	0.047	0.062	12.844
22,000	0.752	0.040	0.053	15.029
23,000	0.752	0.034	0.045	17.584
24,000	0.752	0.029	0.038	20.566
25,000	0.752	0.025	0.033	24.045
26,000	0.752	0.021	0.028	28.104
27,000	0.752	0.018	0.024	32.843
28,000	0.752	0.015	0.021	38.376
29,000	0.752	0.013	0.018	44.840
30,000	0.752	0.011	0.015	52.393

TABLE 6 SPECIFIC HEAT DATA

$\gamma = 1.4$		$\gamma = 1.333$	
Gas	c_p (kJ/kg K)	Gas	c_p (kJ/kg K)
Air	1.006*	CO_2	0.808
N_2	1.041	H_2O (Superheated)	1.921†
O_2	0.917	Hot combustion products in gas turbine	1.150
H_2	14.425		
CO	1.042		

* This value of c_p is 5% low at 650°C and 15% low at 1900°C

† The approximation that superheated steam behaves as a perfect gas is in general poor, but it does hold reasonably well along an isentropic. The value of c_p given is for use in isentropic calculations. (Ref. Horlock, 1959, Proc. Instn. Mech. Engrs., Lond., 173, 779).

REFERENCES

1 Gibbings, J C. Thermo-Mechanics. Pergamon Press, Oxford. (1970)

2 Montgomery, S R. Second Law of Thermodynamics. Pergamon Press, Oxford. (1966)

3 Zemansky, M W. Heat and Thermodynamics. Fifth Edition, McGraw-Hill Book Company. (1968)

4 Lighthill, M J. The response of laminar skin friction and heat transfer to fluctuations in the stream velocity. Proc Roy Soc, A 224. (1954)

5 Shapiro, A H. The Dynamics and Thermodynamics of Compressible Fluid Flow. Vol 1. The Ronald Press Company. (1953)

6 Owczarek, J A. Fundamentals of Gas Dynamics. International Textbook Company. (1964)

7 Liepmann, H W and Roshko, A. Elements of Gas Dynamics. John Wiley and Sons Inc. (1957)

8 Duncan, W J, Thom, A S and Young, A D. Mechanics of fluids, Edward Arnold. (1960)

9 Shapiro, A H. The Dynamics and Thermodynamics of Compressible Fluid Flow. Vol II. The Ronald Press Company. (1954)

10 Strehlow, R A. Fundamentals of Combustion. International Textbook Company. (1968)

11 Shapiro, A H and Hawthorne, W R. The Mechanics and Thermodynamics of Steady One-Dimensional Gas Flow. J Appl Mechanics, Vol 14, No 4. (1947)

12 Kay, J M and Nedderman, R M. Fluid Mechanics and Heat Transfer, Cambridge University Press. (1974)

INDEX